The Fabulous Ford Tri-Motors

Other Books in the Flying Classics Series

Pan American's Ocean Clippers *by Barry Taylor*
The Ercoupe *by Stanley G. Thomas*
The Boeing 247 *by Henry M. Holden*
Beechcraft Staggerwing *by Peter Berry*
The Luscombes *by Stanley G. Thomas*
The Douglas DC-3 *by Henry M. Holden*

The Fabulous Ford Tri-Motors

Henry M. Holden

TAB AERO

Blue Ridge Summit, PA

FIRST EDITION
FIRST PRINTING

© 1992 by **TAB Books**.
TAB Books is a division of McGraw-Hill, Inc.

Printed in the United States of America. All rights reserved. The publisher takes no responsibility for the use of any of the materials or methods described in this book, nor for the products thereof.

Library of Congress Cataloging-in-Publication Data

Holden, Henry M.
    The fabulous Ford Tri-Motors / by Henry M. Holden.
        p.   cm.
    Includes bibliographical references and index.
    ISBN 0-8306-3916-0 (pbk.)
    1. Ford Three-engined Monoplane (Transport plane)—History.
I. Title.
TL686.F66H64 1992
629.133'340423—dc20   91-30482
                                                                                    CIP

TAB Books offers software for sale. For information and a catalog, please contact TAB Software Department, Blue Ridge Summit, PA 17294-0850.

Acquisitions Editor: Jeff Worsinger
Book Editor: Tracey L. May
Director of Production: Katherine G. Brown
Series Design: Jaclyn J. Boone
Cover Illustration: Larry Selman, Waynesboro, PA.
Cover Design: Rick Holberg, York, PA.   FCS

"Of course, one of my biggest interests in the airplane is that it may prove to be the destroyer of war. I think that it will be its biggest service. . . ."

Henry Ford
October 1925

"When I dipt into the future, far as human eye could see, saw the vision of the world, and all the wonders that could be. Saw the heavens filled with commerce, argosies of magic sails, pilots of the purple twilight, dropping down with costly bales. Heard the heavens fill with shoutings, and there rained a ghastly dew, from the nations airy Navies grappling in the central blue. Till the war drum throbbed no longer, and the battle flags were furled, In the Parliament of Man, The Federation of the World."

Alfred Lord Tennyson

Contents

Acknowledgments xi

Introduction xiii

1 The fantasy   1
The Wright brothers   2
The first scheduled airline   6
War clouds   10
William Bushnell Stout   14
The Bat Wing   17
All-metal construction   18
Ford puts out feelers   20

2 The Air Sedan   23
A unique fund-raising campaign   23
The Stout Metal Airplane Company   25
2-AT roll-out   27
Ford Airport   30
Ford makes it official   31
A milestone in aviation   33

3 The Kelly Bill   37
The dangers of flying   40
Stout-Ford merger   42

The short-lived 2-AT 43
A mail route in Ford's future 44
The Ford reliability tours 45
Reliability tours—a significant boost for aviation 47
First commercial airmail contract 48

4 The birth of the trimotor 51
3-AT an ugly albatross 53
A mysterious fire 54
The Ford Flivver 57
The Towle Tri-Motor 58

5 Launching the Tin Goose 63
Ford advertising 69
Construction 70
Molding public opinion 71
Comfort 72
Simplicity 73
First airplane assembly line 78
Stressed skin 78

6 Slow growth 81
Executive Ford Tri-Motors 84
Stout Air Service 84
Charles Lindbergh 86
The 5-AT Tri-Motor 87
Supply and demand 89
The stock market crash 90

7 The race for the coast 93
Crash fever 93
Coast to coast in 24 hours 94
The window that opened and closed at 10,000 feet 95
Discomfort 96
Competition for passengers 98
Ford crashes 100

8 Ford legal problems — 105

The flying pioneer 107
Radio communications 107
Other uses 111
Texaco No. 1 113
Other Fords 115
Diesel-powered Ford 116
The military 117
The Ford Tri-Motor as an investment 117
End of an era 120
A Ford Tri-Motor to Kansas City 121
The first modern commercial airliner 121

9 The flights of the Phoenix — 127

The Tanganyika Star gets camouflage 128
Reflections in a sea of green 129
Island Airlines 132

10 Other Fords — 139

The last of the Tri-Motors 141
Other trimotors 143
Nostalgia 144
Ford the movie star 148

Epilogue 149

A Ford aviation milestones — 153

B Ford Tri-Motor survivors —July 1990 — 155

Endnotes 159

Bibliography 165

Index 169

Acknowledgments

There are several people to whom I am deeply indebted for their generous assistance in helping me document one of the most important aircraft in commercial aviation history. William T. Larkins, with whom I have corresponded for 10 years, provided invaluable help making this an accurate reflection of the history of the Ford Tri-Motor. Over the years he has generously shared with me much of his research and personal photographs, some of which appear in this book. Without those photos this book would not be nearly as exciting.

Penn Stohr grew up under the wing of a Ford Tri-Motor owned by Johnson Flying Service, and today he flies one of the remaining Fords. It was Penn to whom I also turned when I needed in-flight photos of the old Tin Goose. Penn responded generously and with quality photographs, as this book illustrates. Because of space limitations, I must also apologize to Penn for not being able to tell the story of Johnson Flying Services.

A third person who responded quickly and "flew the extra mile" for me was Melissa Keiser of the NASM. Her on-the-scene research when I could not get down to Washington saved me untold time and expense, and her selection of photos was perfect.

Two ladies in the Florida State Photo Archives, in Tallahassee, also deserve special mention. Jody Norman and Joan Morris gave me invaluable assistance and allowed me to rummage through their photo collection until I found just the right photographs.

Others I'd like to thank are Natalie Asseo, of Lufthansa German Airlines, Marilyn Phipps, at the Boeing Airplane Company, and my friend Harry Gann, at McDonnell Douglas.

There is also my research assistant Carole Sandhovel, who has remained dedicated to my projects and did the typing on my first draft.

Last but certainly not least, I would like to thank my strongest supporters: my wife Nancy, my son Stephen, who has become my front-line editor, and my son Scott. They always understood my mission.

Introduction

Today when we fly, we think of the aircraft in terms of a number, usually preceded by a name, e.g., Boeing 727, Lockheed L1011, or the McDonnell-Douglas DC-10. Nowhere in an airport or on the planes will we see the name *Ford*.

Over 75 years ago, a man had a dream, or perhaps a better word would be a vision. This man foresaw the day when vehicles faster than an automobile would transport people and do so in a safe, comfortable, and reliable way. He dreamed of a commercial airline network spanning the United States and the world. His name was Henry Ford, and this is the story of the plane that brought Henry Ford's vision to reality—the Ford Tri-Motor.

We all know Henry Ford's contribution to America. He put America on wheels. With his famous assembly-line concept, he made it possible for thousands of Americans to afford automobile travel. As Ford's Model-T, called the Tin Lizzie, came off the production lines, businessmen, entrepreneurs, farmers, and factory workers all began to buy this new horseless carriage. It reached the hands of the average American, who then used it to change the face of America. Paved roads and highways began to connect small towns with larger ones, and Americans were on the move.

Henry Ford made another contribution to America in the 1920s. It is not as well known, but its far-reaching effects have equaled if not surpassed the achievement of the automobile. He took the fledgling airplane, considered by most people at the time to be a noisy, bothersome, and dangerous machine belonging to the Barnstormer, and transformed it into a convenient way to travel. He was responsible for the beginning of an airline network linking cities all over the United States. He inaugurated the first regularly scheduled airmail service in the U.S. operated by a private company. He started the first air-freight service, he was the first to employ flight attendants, and he was the first to build a paved runway at an airport. He accomplished this and more. He also built a radical airplane design; an all-metal airplane called the Ford Tri-Motor, affectionately dubbed the Tin Goose. The first all-metal, multiengined transport in the United States, the Tin Goose was standard equipment on the U.S. airlines during the late twenties, and until the first flight of the Boeing 247 in February 1933, deserved the reputation of being the only genuine all-metal commercial transport in the country.

The story of the Ford Tri-Motor does not start with an all-metal trimotor, and it does not represent the design of Henry Ford or any single individual. The Ford Tri-Motor, like many other successful airplanes, came about as a result of being in the right place at the right time. There were also many influences and circumstances that made the Ford Tri-Motor inevitable.

The impact of the Ford Motor Company entering the new and poorly managed aviation industry was enormous. The extensive advertising campaign by Ford, the advances in airport design, and the pioneering of airplane assembly-line techniques all had major influences on the air transport industry in the United States between 1925 and 1933. There were other innovative airplane designs, as you will see, but the impact of Ford and his Tri-Motor is one of the milestones in the history of American commercial aviation.

1

The fantasy

Does anyone understand the process of evolution completely? When a living organism reaches a point where physical change seems necessary to continue or better its existence, nature allows the change. Within hours a caterpillar will emerge from its cramped quarters, unfold wet wings, and become a beautiful butterfly. If we look at human evolution against the backdrop of human intellectual development, human beings have progressed from the cocoon stage of the caterpillar to the flying stage of the butterfly in the same relative time. For thousands of years Homo Sapiens roamed the earth, but it is only in the last 90 years that we have conquered the mysteries of flight. Since the human form stood upright, we have dreamed of flying like the birds. The pyramids show the Egyptian god, Khensil, with wings, and the Greek god of commerce and escort to the underworld, Hermes, has wings on his feet. The Roman equivalent, Mercury, sprouts wings from his helmet. The drawings and sculptures of the ancient Babylonians, Assyrians, and Hittites show variations of man-eagles with wings. In 1898, the tomb of Pa-di-Imen (Egypt 200 B.C.) yielded a sculpture of a bird made of sycamore. The bird resembled, in a general way, a Douglas DC-4 with careful attention to the airfoil shape of the wings and tail. So exacting was the detail that archaeologists believe the Egyptians used it to instruct in the theory of flight.

The 14th century saw the first limited application of scientific research to investigate the mysteries of flight. Leonardo da Vinci's drawings were the seeds of man's eventual victory over gravity. That victory, still, would take five more centuries.

The French literary hero of the 17th century, Cyrano de Bergerac, often speculated on manned flight. Jules Verne gave the 19th century a somewhat realistic look at his science fiction of the future.

The Wright brothers

In 1903, when the Wright brothers first conquered flight in a heavier-than-air machine, many people did not believe their accomplishment. There was no fanfare at Kitty Hawk, North Carolina, on that cold December day, just a handful of spectators. Official skepticism of the event, the Wright brothers' cloak of secrecy, and lack of public confirmation added doubt that man had conquered powered flight. A group of people in the Wrights' hometown of Dayton even formed a club called "The Man Can't Fly Club."[1] Most people did not think the Wrights had conquered flight, and the press ignored the event.

In all the world only seven men raised their voices to cheer—the seven on the dunes of Kitty Hawk. On December 17, the Wright *Flyer*, little more than a box kite with an engine, built on years of hard work and genius, opened a new dimension to the world.

On the fourth flight that day, a burst of wind caught the machine as Orville was landing. The *Flyer* rocked and hit a spectator, who fell into the machine and got trapped. Then the wind died down, and the plane stopped rolling. It released its unwilling passenger on the sand, bruised, but alive. The gremlins had tried hard to discourage man from flying, but they had failed.

Accurate stories related by the people at Kitty Hawk were, surprisingly, ignored. Kitty Hawk was far from the news beats, and the Press received no more information than was in the hastily written telegram from Orville to his father in Dayton, Ohio. The telegram announced that his two sons had made four successful manned flights, all against a stiff 21-mile-per-hour wind. The average speed of the *Flyer* was 31 miles per hour, and the longest flight was 59 seconds.

In January 1904, the brothers attempted to clarify the misinformation about their airplane and the Kitty Hawk flight. The gesture had little effect in reversing the general opinion that they were no more than bicycle makers bent on killing themselves in flying machines that did not fly.

By October 1905, the Wright *Flyer III* had covered 24 miles in 36 minutes, 3 seconds. The Wrights decided it was time to invite the Press to witness their success against gravity; but bad weather and engine trouble grounded the *Flyer* for this demonstration. The Press abandoned the Wrights, saying that both the Wrights and the Press were wasting their time. The lack of attention from the Press also discouraged the government from supporting the Wrights. Since the Wright brothers could not obtain patent approval or support from the government, they stopped experimenting for two and a half years, to prevent industrial sabotage.

Toward the end of the first decade of flight, when people began to see airplanes, they associated aeronautical development with people like the Wrights, Glenn Curtiss, and Glenn Martin, in America. In Europe, men like

Louis Bleriot, Hugo Junkers, and Anthony Fokker were the first aviation pioneers.

After World War I, names like Boeing and Douglas emerged as aviation pioneers. These emerging pioneers were young men, just starting out with little experience in business and no experience in aviation.

Another name emerging in the fledgling field of aeronautics was Ford. Henry Ford put America on wheels, but he did not invent the automobile. That credit goes to a Frenchman, Nicholas Cugnot, who in 1769 made the first "road wagon." It was a three-wheeled, self-propelled, steam-powered vehicle. In 1875, another Frenchman, Etienne Lenoir, produced the first carriage-type vehicle using an internal combustion engine. Henry Ford gave the idea of the automobile a practical value by putting it into mass production. Will Rogers, the famous humorist, once said of Ford, "It will take a hundred years to tell whether he helped us or hurt us, but he certainly didn't leave us where he found us."

Ford's success with the automobile may have overshadowed another Ford contribution, one that now has an obscure place in history. Henry Ford had much to do with growing America's commercial wings. He would use the same mass-production idea to give commercial aviation its foundation. Ford set out to prove that commercial aviation was practical and did so with the first all-metal, multiengine, commercial airplane, the Ford Tri-Motor, sometimes called the Tin Goose.[2]

Over the years, there have been several books written on the history of the Ford Motor Company. None of these books contains much on Ford's early aviation efforts or on the Ford Tri-Motor. Some books gloss over this important era in both the Ford history and aviation history with barely more than a paragraph or two. The last book on Ford, published in 1989, has two brief references to the Ford Tri-Motor.

When Ford's Tri-Motor arrived, commercial air travel was not popular. In fact, it was downright dangerous. Ford used his corporate name, extensive advertising, and publicity campaigns focused on safety to gain commercial acceptance of air travel. Thousands of people had their first plane ride in a Ford Tri-Motor, and the foundation of the passenger airline system in America started around this plane. The Ford Tri-Motor rattled and shook and made a deafening noise to those aboard, but it was a quantum leap in technology over the other airplanes of the day.

Air travel in America was inevitable, but Ford accelerated its acceptance by years. The 200 Tri-Motors left a legacy of accomplishment and an example for the future.[3]

Over 100 airlines have flown the Ford Tri-Motor in North America, Central America, South America, Europe, Australia, and China.[4] (Figs. 1-1, 1-2, 1-3, 1-4). Dozens of private companies used the Ford Tri-Motor for executive

1-1 An Eastern Air Transport (later Eastern Airlines) Ford Tri-Motor.

1-2 A Pan American Airways (later Pan Am) Ford Tri-Motor.

1-3 This 1970 photograph of a Transcontinental Air Transport (later TWA) Tri-Motor was taken at Harrah's in Reno, Nevada. Note the rare combination of a modern tank and the Tri-Motor (5-AT-8).

4 The Fabulous Ford Tri-Motors

1-4 An America Airways Ford (later American Airlines).

transportation, sales, promotions, and transportation of freight. Henry Ford was the first to use his airplane for all these purposes.

As big and boxy as the Ford was, pilots looped it and flew it through snap rolls, Dutch rolls, and lazy eights, not a bad repertoire for a 14-passenger airliner. It did not take long for people to discover its remarkable ability as a heavy-duty freight carrier. When it came to hauling freight, the Ford Tri-Motor surpassed every other prewar American commercial transport, except the Douglas DC-3.

During World War II, Henry Ford made a name for himself in aviation with his 50,000-plane production line, and the famous four-engine Liberator bomber production at Willow Run. His initial entry into aviation in the 1920s, now obscured by history, is his real legacy. His money, dedicated vision, and mission irrevocably launched America into the air age.

The Ford interest in aviation literally grew with the development of the airplane. Henry Ford, a pragmatic individual, did not give much thought to aviation until 1909, when his 16-year-old son, Edsel, began building an airplane modeled after a French Bleriot monoplane. Young aviation pioneers like Glenn Curtiss wanted Ford and his money in aviation, but Ford resisted getting involved until Edsel and his friend Charles Van Auken came to him for help.[5] Ford was happy to see his son involved in something mechanical and rented a barn for the budding young aviators to build their machine. He even provided many machined parts from his Detroit factory. The plane, called the *Auken-Ford Flying Machine*, had a 28 hp direct-drive Model "T" engine. The plane was aerodynamically inefficient, unstable, and a mixture of originality and early Wright and Bleriot designs. It was an unusual craft for its day.

The fantasy 5

It had a frame of metal tubing, wings of silk, and a tricycle landing gear (made of bicycle wheels), an idea ahead of its time. The pilot wore a yoke around his shoulders, and when he shifted his weight to the left or right, he moved the corresponding aileron (a principle used earlier by the Wright brothers).[6] Unlike the Wright machine, where the pilot lay prone in the fuselage, the pilot sat upright as in the Bleroit design. The plane lacked power, so Henry gave the boys a souped-up Model "T" engine. With that engine, Van Auken made several successful flights (Edsel never flew the plane). On one flight, it crashed into a tree, but Van Auken escaped serious injury. After Van Auken's close call, Ford decided not to help the boys rebuild the plane, considering its marginal performance. There is speculation that the success of the Wright machine in the Fort Meyers air trials put a damper on Ford's ideas of building airplanes.

The first decade of American aviation saw little design progress. Some daring Americans were taking to the air and many were dying in the attempt. Most people did not take flying seriously (Figs. 1-5, 1-6, 1-7).

When war in Europe broke out in 1914, America in many respects had not yet entered the 20th century. The tiny American aviation industry had a total of 168 wage earners who netted a base of $135,000. They worked for 16 aircraft companies, producing a total of 49 military and civilian airplanes, valued at $750,000.[7]

The first scheduled airline

The first hint that the airplane had possibilities as a commercial vehicle in the United States came a decade after the Wright brothers' first flight. On January 1, 1914, the first scheduled airline, the St. Petersburg-Tampa Airboat Line, carried a man across Tampa Bay.

Paul Fansler, an electrical engineer, founded the airline and is perhaps America's first airline promoter. On December 17, 1913, the 10th anniversary of the Wright brothers' first flight, Fansler signed a three-month contract with Tampa city officials. He agreed to allow the mayor of Tampa, A.C. Phiel, to become the airline's first passenger for a tidy sum of $400. (Phiel was the highest bidder at the auction for the first ticket.)

The plane used by the airline was a Benoist (pronounced Ben Wa) Type XIV, made in St. Louis, Missouri. It had a 75 hp Roberts engine, a top speed of 64 mph (about $1/10$ the speed of a modern jet), and a wing span of 35 feet (about $1/6$ the span of today's modern jets) (Fig. 1-8).

The businessmen of Tampa agreed to subsidize the airline up to $1,200 in the event the public lacked the courage necessary to make the 18-mile, 23-minute trip over water. Surprisingly, the subsidy was not necessary, and the only cost to the city of Tampa was $250 for the erection of a hangar.[8]

1-5 Harriet Quimby was the first woman to earn her pilot's license.

1-6 Others like Ruth Law seen here in her Curtiss Pusher Loop Model, an early aerobatic plane shown here at Daytona Beach, were lucky. At the height of her career in 1921, while making $9,000 a week, she gave up flying and lived for 87 years.

1-7 The U.S. Navy's first plane was little more than sticks and wire. The Triad, built in 1911, was to be either a land or a seaplane.

8 The Fabulous Ford Tri-Motors

1-8 The St. Petersburg-Tampa Airboat Line used one Benoist Type XIV for four months before economics shut it down.

The line carried everything from people to hams and chickens, and charged $5 per hundred pounds for people. The line flew without incident, and in the 90 days logged on the books, it lost only eight days to weather or mechanical problems.

When the three-month contract expired, the airline continued operations. The threat of a Mexican War, the European war, and the tightening of financial conditions in the United States closed it down a month later. When the line shut down, Fansler disappeared from the aviation scene.

It was not until 1934 that Tampa or St. Petersburg had another scheduled airline service. That year, National Airlines began service from the two cities to Daytona Beach. In the first year of that operation, they carried 400 passengers compared to 1,204 passengers carried in three months on the 1914 Airboat Line[9] (Fig. 1-9).

Early aviation exacted a terrible toll on its participants. In an ironic twist of fate, the men who played a major role in forming the country's first scheduled airline were dead before the start of the second airline. Pilot Tony Jannus, who made the first flight, lost his life over the Black Sea in 1918. His brother Roger, who helped with the airline, crashed his aircraft on the Western Front during World War I and died of his injuries. Thomas Benoist died in an automobile accident in 1917.

The fantasy

1-9 The Wright Model B is shown here over the city of Tampa in 1914. The pilot is Ruth Law.

War clouds

War clouds on America's horizon sparked Ford's renewed interest in the developing science of aeronautics. Ford believed that aviation should be kept out of the theater of war. The Wright brothers had a vision of the possible sequence of aviation development: first, use airplanes for military reconnaissance and exploration; then, after further development, would come the transportation of passengers and freight. The Wrights, like Ford, believed the airplane might prevent war by creating an awesome military deterrent.[10] Ford once called the airplane "one of those damned war-making machines."[11]

To exercise this deterrent, Henry Ford proposed the construction of 150,000 airplanes. Ford had ideas of using his moving-assembly-line tech-

10   The Fabulous Ford Tri-Motors

niques to mass-produce airplanes. Edsel Ford had asked the Ford Motor Company of England to supply an efficient English airplane to serve as a model for production. Unfortunately, the U.S. government never gave their support to this proposal, and Ford did not begin production of the airplanes. Ford, like the other auto manufacturers, switched his manufacturing ability to building the famous Liberty engine. (Ford produced 3,950 Liberty engines for the war effort.)[12] This engine later powered the American-designed NC-4 planes in the first Atlantic crossing and other historic American flights of the postwar period.

During the war, Ford's factories installed the Liberty engines in Curtiss JN-4 Jennies (Fig. 1-10) and American-made, British-designed de Havilland DH-4 airplanes. With the infusion of government money, a new industry developed overnight to address America's aviation needs.

1-10 The Curtiss Jenny was a popular plane for airmail and police departments. Here are three Jennies belonging to the Hialeah Police Department.

On May 24, 1917, the French government requested that the United States furnish 4,500 war planes for the upcoming Spring 1918 offensive. The United States also projected a need for an additional 17,000 aircraft.[13] At peak production in 1918, there were over 14,000 planes and 31 companies producing aircraft parts valued at $14 million.[14]

The aircraft industry prospered and grew during the war, but the armistice suddenly put the industry in danger of extinction. Within days of the armistice, the government canceled orders for 13,000 aircraft and 20,000 engines. From April 1917 to November 1918, American aircraft producers delivered 13,984 planes and at the time of the armistice had the capacity to turn out 21,000 aircraft per year.[15] The sudden end to the hostilities caused the huge surplus of airplanes and engines to glut the market. The end of the war choked off the lucrative government contracts, and the once-growing industry manufactured only 780 planes in 1919, with all but eight for the military. In 1919, Edsel Ford had the Ford Motor Company revise its corporate charter, adding a provision for aircraft production.[16] Perhaps he had vision or had already heard of William Boeing's progress. Boeing, a lumber company

The fantasy

owner, had built about 50 training planes for the Army and was looking into the possibilities of commercial aviation (Fig. 1-11).

After the war, some farsighted companies, like the American Railway Express Company, began exploring the commercial possibilities for the airplane. In 1919, they flew a converted Handley-Page bomber on an experimental flight carrying 1,100 pounds of cargo from New York to Chicago. A storm grounded the plane on a racetrack outside Pittsburgh, so they put the cargo on a train to finish its journey to Chicago. The experiment failed, but from it came the idea of combining air and rail for the transportation of freight.[17] The United States, however, was not ready for commercial aviation. Too many parts of the equation were missing. All-weather planes, flight instruments, structurally sound aircraft, and the economics to pay a profit were all yet unborn and still more than a decade away. It was not until 1927 that National Air Transport and Colonial Air Transport first offered organized service in the East. Boeing Air Transport and Western Air Express offered similar service in the West. Seven years later, commercial aviation would finally come of age and become a money maker, with the introduction of the Douglas DC-3 (Figs. 1-12, 1-13).

There were many people who also questioned the practical use of the airplane to attract commerce. Most designers thought that further development of military machines was necessary to arrive at an efficient commercial airplane design. Many also questioned the future of this new means of transportation. One aviation pioneer, Donald W. Douglas, recognized the potential of commercial air transport. "Speed is the most outstanding present-day

1-11 William Boeing's first plane, the B&W.

12 The Fabulous Ford Tri-Motors

1-12 The famous Douglas DC-3, the plane that changed the world and quickly replaced the lumbering Fords.

1-13 This Pan American DC-3 seen here in its prime in 1940.

advantage of the airplane," he said. "I rank passenger carrying first in importance. Many of our personal and business fares can be successfully conducted only by a personal meeting. Correspondence or the telephone cannot supply the complete satisfaction of actual personal contact. Airplane builders must come forth as commercial operators and take the risk. One way or another

The fantasy 13

they must carry passengers on schedule, comfortably and without a mishap for a reasonable period."[18]

At this point, surface transportation provided the most efficient and inexpensive means of carrying freight. "There have been incidents," Douglas said, "where a certain business will spring up to form the rapid transportation of perishable goods, like medical supplies, motion picture films, and luxury items. Where speed is of the very quintessence of transportation, the airplane will have a very definite field and a profitable one."[19]

Douglas was predicting the future, but it would take time for aviation to find its way through the uncharted heavens, the morass of primitive technology, and bureaucratic indifference.

In England, commercial aviation was not standing still. By 1920, England had flown 4,200 people and 57,000 pounds of freight 84,428 miles, without a single injury or loss of goods. In the United States, steady commercial service had yet to begin, although the Post Office was pioneering Air Mail Service.[20] As early as 1910, someone in Congress introduced a bill to consider airmail operations. The *New York Telegraph* found the idea preposterous. "Love letters will be carried in a rose-pink airplane steered with Cupid's wings and operated by perfumed gasoline," was its patronizing and sarcastic reply.

The first two decades of powered flight saw continued but sporadic efforts to launch commercial aviation in the United States. There was a brief attempt by a passenger airline flying between New York City and Atlantic City. The company used Aeromarine flying boats, but the combination of bad flying weather, rough landings, and mechanical trouble forced them out of business within three months.

Alfred W. Larson, another aviation pioneer, also tried making a go of commercial aviation with his C-2 *Larson Liner*. This multiengined behemoth was as big as the future Douglas DC-3, but so uncomfortable, noisy, and dangerous that it did not last. The plane eventually became a hamburger stand in Milwaukee.[21] His second effort, the *L-4* airliner, crashed on its initial takeoff. His planes were the first multiengined planes designed and built as passenger planes in the United States.[22]

There was no encouragement coming from the government and little from the private sector. The *Scientific American*, an early proponent of commercial aviation, said, "American aviation must depend on the commercial side of flying. Scheduled regular service is the only type of flying that can bring peacetime prosperity to our aviation industry." The editors knew that air transportation would one day influence commerce in the vast landmass called the United States. "Today," they said, "we measure distance in the time it takes to go from one point to another. Citing a distance of 1,000 miles between New York, and Chicago makes no impression on a businessman.

He only knows about the twenty hours by train. By aeroplane the two cities would be 7-8 hours apart."[23]

In 1921, the *New York Evening Post* declared, "Public interest in the use of airplanes for commerce in this country has not yet reached the point where the industry can be supported on a regular commercial basis."[24] They were correct, and by 1923, two decades after the Wright brothers first successful heavier-than-air flight, there were still no organized airlines. Commercial aviation in the United States consisted mostly of air taxi flights and sightseeing tours from bumpy corn fields at county fairs. Investors were wary of an enterprise that promised little or no return for years to come. They were reluctant to invest in an enterprise where the investment frequently crashed and went up in flames. Aviation in the United States would have to show maturity, responsibility, and the capacity for commercial service in the interest of man. Above all, it would have to prove it was safe. This would not be an easy job.

William Bushnell Stout

During this confusing period in postwar America, the activities of creative men emerged to bring the organized air age to America. One of those men was William B. Stout, regarded by many as eccentric, but by all as a brilliant, creative inventor. His engineering skills and readiness to try new ideas earned him the nickname "The Socratic Gadfly of Aviation."[25] Born in Quincy, Illinois, in 1880, Stout was the grandson of David Bushnell, an early inventor-genius. It was Bushnell who, during the Revolutionary War, built the first workable submarine.

Stout was also an accomplished inventor-engineer. At the age of 14, he designed and built his first model airplane. It was made of tissue paper and cork, with two propellers of chicken feathers. The feathers revolved in opposite directions, driven by the two twisted rubber bands. Surprising at the time to the young Stout, it flew.[26] He tried to improve the model by substituting turkey feathers, but it did not fly. So prolific an inventor was Stout that he would go on to own more technical patents than any other American, except Thomas Edison.[27]

Stout dropped out of college in his freshman year because of his poor eyesight and was not a graduate engineer. By 1917, Stout had worked in a variety of professions that included teaching and journalism, and a job in 1912 as the editor of *Aerial Age*, an early aviation magazine. He also wrote for the *Chicago Tribune*, lectured on artificial flight, did advertising, selling, automotive design, and finally aeronautical engineering. Unlike many gifted inventors, he also possessed a talent for promoting his inventions in unconventional and often very successful ways.

At the outbreak of World War I, Stout was chief engineer of the Aircraft Division of the Packard Motor Car Company. His job was to match the new Liberty engine to the wood and wire devices called biplanes (Fig. 1-14). Soon afterward he went to Washington to serve on the newly created wartime Aircraft Board.[28] Stout's job was to accelerate production of war planes. He soon discovered the state of American aviation design was unoriginal, flimsy, and dangerous. The planes were slow, lacked brakes, and had only a rudder to steer with on the ground. Stout also read the poor performance reports on the Army de Havillands. With a 200 hp engine, 93 hp went into lift and 107 hp went into fighting parasitic drag. To further compound his problem, the new 420 hp Liberty engine installed in a Curtiss biplane, previously powered by a 200 hp engine, increased the plane's speed only 5 mph. He knew an invisible force was robbing the plane of speed and efficiency. He set to work to remedy this problem.

America desperately needed aircraft during World War I. Henry Ford had not gone ahead with his wartime proposal to build 150,000 planes, and American aircraft design lagged far behind European progress. Americans copied English and French planes during the war, and in this humiliating atmosphere, Stout's creative energies were soon evident.

1-14 This cross section of a Jenny shows the flimsy workmanship.

16 The Fabulous Ford Tri-Motors

The Bat Wing

The Army was wrestling with the problem of parasitic drag at McCook Field, near Dayton, Ohio. After hundreds of hours of flight testing, they discovered that a small wire between the wings, vibrating from the wind sweeping over it, caused more drag than an inch-and-a-half thick solid wooden strut. Stout calculated a biplane would need 300 hp to pull it through the air, leaving 100 hp for lift. Like da Vinci, Stout studied the birds, and from his observations, he reasoned that the key to speed was monoplane construction. After all, birds did not need two sets of wings to fly. He proposed a radical airplane design that would use a single thick wing with internal bracing. His design would eliminate the struts and wires of the conventional biplane and theoretically create a more efficient airplane. Unknown to Stout at the time, Hugo Junkers in Germany was also working on a similar design. The engineering drawings of Stout's design impressed the Aircraft Board, and they awarded him a contract to build a mockup of his new plane. Stout then moved to Detroit to supervise construction.

Not content with such an unimaginative task, Stout decided to build a full size flying model of the plane. The Bat Wing was essentially a flying wing, a design that the industry would not get around to recognizing for almost 30 years. It had a thick wing, with an inverted delta shape extending back to the tail surfaces. It was unique in structural material and design. Stout, aided by a newly developed casein glue, had built the first plywood veneer used as an airplane skin in the United States. This design, mated to the internally braced cantilever wing, had never before been seen in the United States or Europe.[29]

Stout took the Bat Wing to McCook Field for its air trials, and the Army installed a 150 hp Hispano Suizza engine in the craft. Before the plane could lift off, a water pump shaft broke, creating enough steam pressure to blow the radiator open. The Army repaired the plane and filled the radiator with ice water, and the test pilot managed to get the craft into the air. Again the plane overheated, and the pilot made an emergency landing. The short flight did prove that the Stout design was feasible and that there was a possible future for thick-winged monoplanes.

Unfortunately, the Armistice arrested aviation development in the United States, and the government lost interest in Stout's experiments. Stout's financial backing to bring his design to a practical stage dried up overnight. For Stout and other aviation designers, the next few years would bring frustration and discouragement. Grover Loening, another aviation pioneer, said, "In a moment I realized that this great opportunity was gone forever. I should not say I was bitterly disappointed, because we all wanted the war to end, but the truth is it was a terrible blow."[30]

A few American-designed airplanes did come to the practical flying stage by the time the war was over. One was the Martin MB-1. Production of the Martin MB-1 bomber came too late for its participation in actual combat, and at the time of the Armistice Martin had orders for more than a dozen planes.

When peace came to America, there was little use for the bomber and Martin advertised it in the industry's leading journal, *Aviation* Magazine. The ad called it "the Most Important Aerial Development of the War. Officially it has surpassed the performance of every competitor. It is the forerunner of the aerial freighter and twelve-passenger airplane."

The editor commented on the design: "The Martin twin-engine bomber constitutes an important development in bombing airplanes of original American design. In its official tests it easily surpassed the record of a similar bomber, either here or abroad. The machine shows excellent workmanship, and such thoroughness of engineering that we complement the Martin organization on the talent of their personnel."

Donald Douglas, the plane's designer, in a feature article said, "The general characteristics that make the machine stand out from the others of this type are its flying qualities and high efficiency. These serve to assure us that the utility and practicability of large machines for both civil and military use have been definitely realized." Unlike the editor, Douglas put civilian aviation before military use.

After the dubious success of the Bat Wing and the commercial hype the Martin bomber was getting, Stout returned to Detroit convinced more than ever that commercial aviation had possibilities. Robert Stranahan, president of the Champion Spark Plug Company, understood Stout's theories and recognized that without government support Stout had little chance of his design becoming a reality. He agreed to finance a $15,000 commercial Bat Wing, a four-place cabin plane that used the newly developed plywood finish. This plane, instead of being powered by the 150 hp Hispano Suizza engine, had a 200 hp, eight-cylinder Packard engine. Stranahan used a bit of political muscle and the Navy took a second look at Stout's Bat Wing. With a larger engine, the plane did slightly better the second time. After the official test flight, the Navy bought the Bat Wing. Unfortunately, the plane was barely aerodynamic, and severely underpowered for its weight. The tail surfaces next to and behind the huge wing chord blanked out in landing and turns. Much more engineering research was needed, but it did prove beyond question the feasibility of the Stout thick-wing design. It carried a 1,170-pound payload to 5,000 feet.[31]

All-metal construction

During the construction of the Bat Wing, Stout discovered that the Aluminum Company of America had a new alloy with a tensile strength nearly double that of ordinary aluminum. He believed that the strength and lighter weight of this metal would be ideal for airplane construction. The Navy was also beginning to show an interest in the possibilities of metal for airplane construction. Stout, in his unique sales style, convinced the Navy that he should get the $50,000 contract to develop a twin-engined, all-metal torpedo bomber. Like the original Bat Wing, the plane would be completely trussed internally, but would have two 150 hp engines. To reduce drag, it would have retractable radiators in the wings.

Stout still had ahead of him the difficult task of proving that metal could be used efficiently as a structural material in aircraft construction. He knew wood was not the right material. Glue, bolts, and screws caused weak spots. Wood splintered on impact, and it was flammable. On the other hand, engineers could measure the strength of metal and accurately gauge it to within two percent. Metal also did not splinter on impact or burn like wood. To most people in American aviation, Stout was crazy. The idea of an all-metal airplane was beyond them. A metal airplane was just too heavy.

Stout answered his critics with a tongue-in-cheek reply, "Any plane built of wood starts getting a disease after six months," he said. "That disease is veneer-eal disease."[32]

To build this all-metal plane for the Navy, Stout rented a factory in Detroit and brought together a group of young men inexperienced in aviation design. He did so purposely. He did not want them bound by the engineering standards in the industry. George Prudden, a recent graduate of the University of Minnesota and a specialist in reinforced concrete structures, was responsible for stress analysis and developing the new structure; Stanley Knauss, an automobile salesman Stout remembered from his journalism days, became his sales manager; and Glenn Hoppin, an electrical engineer, became Stout's business manager. Stout called his company the Stout Engineering Laboratory.[33] These men, helped by others, began to develop the techniques for fabricating aluminum alloys into the structure of a flying machine. After exhausting research and trial and error engineering, Stout's team finally developed techniques for forming, rolling, and bending the cold metal into the shape of his airplane design. At that point, work went forward quickly on the Navy's torpedo plane. "As our work on the all-metal torpedo plane went forward", Stout said, "we saw very clearly that the contract price of $50,000 would not begin to cover the costs." It was then that his friend Stranahan came to the rescue with more funds.[34]

Stout's ideas were attracting attention in Detroit, and many industrial leaders visited the plant to watch his progress. One of these men was William B. Mayo, the self-taught chief engineer of the Ford Motor Company. One of Mayo's jobs was to report on American aviation developments to Henry Ford.

Ford puts out feelers

In the postwar period, range and carrying capacity of the German dirigibles had impressed Henry Ford. Still anxious to put aviation to peaceful uses, and believing dirigibles would be the air transport vehicles of the future, Ford announced that he would sponsor a dirigible building program. Discussions between Mayo and the Navy Department proved fruitless, but in June of 1920, German officials offered some of their dirigible patents to Ford. Ford instructed Mayo to go to Europe to study the German dirigible industry and other aviation developments on the continent. In the Fall of 1920, Mayo returned to Detroit with a surprising report. Germany was developing a strong dirigible industry, but Europe had also moved forward on the possibilities of the airplane as a passenger and cargo carrier. The report shocked Henry Ford. Communications between Europe and the U.S. were poor, and Europe had gone on to begin weaving the threads of an airline system. The situation was ironic. The United States had pioneered aviation, and while Europe moved ahead, the United States was still in the Barnstorming Era (Fig. 1-15). The European governments had either nationalized the germinating aviation industry or provided heavy subsidies for its growth. By 1921, 10 airlines linked cities in Europe, with five carrying passengers and freight in England.[35]

1-15 Here Mable Cody is shown flying the barnstorming route in Daytona Beach, Florida. One of her most famous tricks was the car-to-plane transfer.

Because of his discoveries in Europe, Mayo watched Stout's activities with a sympathetic eye. He kept progress reports flowing up to the Fords, as Stout struggled with the problem of building an all-metal airplane. When Henry Ford heard of Stout's experiments, he began to see the commercial possibilities of the airplane. He later told a Detroit newspaper man, "The airplane is going to enlarge the work of the automobile. The motorcar has mixed people up so thoroughly that one can hardly fool any American about any part of his country. But, they can be fooled about other parts of the world. The airplane will stop that. When the plane becomes popular, it will put power in the people's hands just as the motor car has. When politicians propose a war, the people will know why, and ask questions. They will make short work of these war-makers."[36]

On May 21, 1922, Stout's first twin-engined, all-metal torpedo plane was ready for its official Navy flight test. Stout's test pilot, Eddie Stinson, had flown the plane a dozen times at speeds of up to 112 miles per hour before the official test. On the day of the official test, the Navy insisted their pilot replace Stinson. The Navy pilot, unfamiliar with the craft and very nervous, crashed while attempting a downwind landing, wrecking the craft beyond repair. Stout described it later as "a perfect three-point landing about 15 feet too low." Stout faced complete destruction of his dreams. The wreck of the torpedo plane meant a loss of $165,000 in funds already paid to Stout.[37] The Navy canceled the contract and stopped funds for further experiments. The Stout Engineering Laboratories was deeply in debt and Stout was down, but not out.

Years later, Stout would describe the next few months as a frustrating attempt to recuperate mentally and financially from his losses. Stout was not one to remain discouraged for long. He decided he was through with government contracts and would not waste the experience and knowledge gained during the construction of the torpedo plane. He stubbornly and creatively turned to the idea of building a commercial plane, one that would use his idea of all-metal construction.

2

The Air Sedan

Stout visualized an enclosed-cabin, monoplane design with four seats, and large enough to carry freight, too. While the Air Sedan began to take shape on the drawing board, the cold facts of economics took over. Stout was broke and had to close the laboratory. A few assistants remained with him, making toys designed by Stout and rarely getting paid. Bowling games, a miniature golf machine, submarines, and other imaginative toy designs came from his small shop. With the royalty checks from these and other Stout-designed toys, his small group remained alive but unable to proceed with his airplane design.

With his goal of a commercial airplane still on the drawing board, Stout went back to Bob Stranahan. His friend advised him to form a new company and seek private funding. Stout had barely $100 in the bank and a typewriter. He could not sell stock to the public, and without collateral, a bank would not loan him money.

A unique fund-raising campaign

The creative Stout decided to launch a fund-raising campaign unique in the history of finance. The approach was simple, but dramatic and effective. He wrote to 100 Detroit industrialists and asked each to invest $1,000 in his company. In return, Stout promised nothing but the opportunity to contribute to the building of an aviation industry in Detroit. Stout emphasized that the undertaking would not be profitable and that they would most likely never see their money again. One investor said, "It was the damndest solicitation I ever heard!"[1] In his letter, Stout promised a newsletter to each subscriber, covering the progress of aviation. Stout drew airplane illustrations, and his wife and daughter traced the drawings with carbon paper onto the newsletters.

At the time, the ordinary working man did not consider commercial aviation a serious investment. The postwar recession further restricted investors. William Boeing was struggling, and to pay the bills, he branched out to manufacture sea sleds, bedroom furniture, and arm chairs. Boeing even looked into the market potential of Ouija boards.[2]

When Stout mailed the letters to the industrialists, he had no idea the response would be so positive. He received 65 replies with checks and requests for future issues of the newsletter.[3] Stout hurried from one prospect to another, and the funds came rolling in. Years later, Stout learned that many recipients read the newsletters carefully and kept them on file for years.

Stout's campaign was successful for several reasons. He played on the industrialists' pride in their city, and also on their guilt. "If Detroit expects to take its place in aviation," he said, "someone here has to build planes." Many of Stout's subscribers recommended others for Stout to visit. Among those Stout would meet was Henry Ford.

During the 1920s, the Ford Motor Company received thousands of letters from would-be inventors and crackpots asking for money to sponsor their inventions. Stout's letter was not unusual; it was the sketch of an airplane that caught Edsel Ford's attention. Edsel did not discard the letter because he associated the author's name with some of his chief engineer's reports. Edsel, too, had more than a passing interest in aviation. He offered Stout more than $1,000, but Stout was adamant. He did not want to show favoritism among his sponsors. Edsel decided to get around Stout's requirement by sending Stout $2,000 ($1,000 of his father's money). Among other Stout subscribers were the country's industrial leaders of the 1920s: the Fisher brothers (car bodies), C. F. Kettering (General Motors), W.B. Mayo (Ford), Marshall Field (department stores), Albert Champion (spark plugs), Horace Dodge, Harvey Firestone (tires), and Walter B. Chrysler. Eventually, 99 men contributed to Stout's campaign.[4] Ford's contribution, without a doubt, was more significant than the others. Other automobile manufacturers subscribed to Stout's newsletter, but Ford had more technical know-how and a larger financial base than the others. Ford was also the nation's third-largest employer, with over 200,000 employees and $1 billion in assets. His name alone gave Stout and aviation an unprecedented boost.

One may ask why these rich and prominent men invested their money in a no-guarantee return. It was not charity; it was not even sound business. To some, no doubt, it was a game and they were sportsmen. Others had seen the wartime aviation industry's growth and felt the investment would pay off.

Perhaps some had an intuitive sense and a faith in the future of aviation. Whatever their reasons, these investors kept Stout going.

With enough funds to restart his company, Stout incorporated the Stout Metal Airplane Company on November 6, 1922, "to develop and manufacture aircraft."[5] Even with an impressive list of sponsors, Stout was not out of the woods. There was rent, utility bills, a monthly payroll of over $400, and other obligations.

The Stout Metal Airplane Company

With the new money, Stout went forward with the Air Sedan. His goal was to build the four-place monoplane with a cantilevered wing and all-metal, corrugated construction. The corrugated skin made a lighter and stronger structure than the flush-metal skin of the torpedo bomber.

Hugo Junkers, one of Germany's leading aircraft designers, was using corrugated metal in his planes. Stout had learned some lessons in metal-plane construction from a Junkers plane he had repaired for Eddie Stinson. Stout immediately saw an advantage in the corrugated metal. Making the plane's fuselage entirely of corrugated metal strengthened the plane. Some critics have said that Stout borrowed on the Junkers design, but many ideas independently developed by Stout and his engineers were superior to the German design. In the final analysis, Stout owed little to Junkers.

On February 17, 1923, the Air Sedan (1-AS) (Fig. 2-1) was ready for its debut. On the test flight, it was painfully obvious that the plane lacked power. The 90 hp OX-5 engine was simply not powerful enough to haul the plane off the ground and maintain stable flight. The more powerful 150 hp Hispano Suizza engine was rare in the United States and expensive, so Stout

2-1 The only Stout AS-1.

had settled for the Curtiss OX-5 engine. In 1923, the engine makers had not yet begun producing in quantity the reliable radial engines. Stout maintained that only three engines had any degree of reliability: the 420 hp Liberty engine, the 90 hp OX-5, and the 150 hp Hispano Suizza.

The resourceful Stout "found" an Hispano Suizza engine, and with this new engine, the plane flew satisfactorily. The flight test proved the soundness of the structure, the wing arrangement, and other engineering details. Even with the more powerful engine, the plane was still of little practical value. The OX-5 was too small for the plane, the Hispano Suizza engine was not available in practical quantities, and the Liberty was too powerful for the Air Sedan. The plane was also too expensive for a personal plane and limited in carrying capacity (three passengers) for commercial use.

By 1923, there were still no aircraft in the United States that could carry a decent load of passengers or cargo any respectable distance. The few commercial planes available were single-engine biplanes, made of wood and fabric, with top speeds of about 100 mph. Most of these were of World War I vintage, like the de Havilland GH-4, a two-seat observation plane.[6]

Henry Ford watched the test flight and was not impressed. Stout told Ford that what he needed was more horsepower and that to get it he would need more money. Ford countered that what he needed was more airplane. Stout's mind was always working, and he saw his opening. He used the opportunity to discuss one of his ideas with Ford. His "new plane" would have smoother lines than the Air Sedan, and be more powerful. It would carry a payload of 10 people (two crew and eight passengers) or the equivalent in cargo, have a high wing, and use the 420 hp Liberty engine; it would be a true commercial airplane. Stout called this, his second commercial design, the Air Pullman. When it carried freight, he called it the Air Truck. Later, Stout combined these two names into the Air Transport (2-AT). Stout took his proposal to his other sponsors, and they unanimously agreed he should begin work on the Air Pullman.

The time was right, Edsel Ford realized, to exploit Stout's genius. The Ford Motor Company should pump some money into his research. Stout had read Edsel Ford like a book and took the advantage. Stout knew the day of money-making aviation would occur, but he could not afford to sit around and wait for it to happen. He wanted to be a part of the process. He asked Edsel Ford for a suitable airfield to make test flights. He pointed out that he had to carry on demonstration flights at an airfield 26 miles outside the city. Edsel said he would consider Stout's request. Aside from the fact that Stout was using Ford money, Edsel Ford's own aviation interests prompted him to seek a more active role in Stout's company. When Edsel approached Stout and asked to subscribe to more stock, Stout obliged. On December 31,

1923, Edsel Ford became a director in Stout's company. Stout immediately called his first board meeting.

Stout had great hopes for his new 2-AT design and asked his directors to visit the plant and drive a symbolic rivet into the plane. "Driving a rivet into this plane," he said, "will mean something to you and your grandchildren."[7] Stout had good reason for his optimism. The plane, based on the Air Sedan design, was progressing rapidly and had the promise of fulfilling Stout's prediction. Stout had put the experience gained in designing the 1-AS to good use.

Stout described the development of the machine: "The plane was designed out of a C.A.A. handbook of about twelve pages, and was in a large degree imaginary. The structure was designed by a man just out of the University of Minnesota, graduating as a specialist on reinforced concrete structures."[8] (Stout was referring to George Prudden, who later went on to Lockheed and established many standards for internal airplane design.)

In building his transport, Stout made many technical innovations. Because his design was far ahead of the other manufacturers, he could not obtain new information from the industry or reliable research sources like the National Advisory Committee for Aeronautics (NACA). Created in 1915 by Act of Congress, the NACA was the first government agency charged with supervising scientific research in aeronautics. It became the National Aeronautics and Space Administration (NASA) in 1958. The NACA published what it had, but as Stout recalled, "The sum of this data could be contained in a pamphlet 1/8 inch thick." Stout did get some information for the wings for the 2-AT from some experimental contours the committee had developed.[9]

2-AT roll-out

On April 23, 1924, the first 2-AT rolled out and fulfilled all Stout's expectations in the test flights (Fig. 2-2). The *Detroit Free Press* heralded the plane as "Detroit's first bid for commercial aviation."

Stout's design was revolutionary. The Air Transport had a place for a co-pilot as well as a pilot. This in itself provoked controversy (other American civilian planes at the time had one pilot). When it became known that Stout intended to place the pilots forward of the wing's leading edge, criticism mounted. Pilots should be back behind the wing, and close to the tail, everyone said. In case of a crash they would have less chance of getting hurt.

Stout had another motive. He was designing his plane to carry passengers, and he felt passengers who were paying for the ride should have the most protection. Stout also felt that if the pilots were up front, they would have better visibility. Stout placed the pilots and cockpit in front of the wing

2-2 One of Stout's 2-ATs at work carrying mail for Florida Airways. Notice the open cockpit and thick wing root.

and passenger cabin. To provide better visibility, he installed a windshield of celluloid, the only transparent plastic available at the time.[10]

Stout named the plane *Maiden Detroit*, and the name served a dual purpose. The first was to promote civic pride and interest. It was also a play on words, a joke Stout played on the uninformed. He was really saying, "Made in Detroit."

Henry Ford liked the new 2-AT, so he ordered several planes from Stout. Ford was going to put aviation to a peaceful use by starting a transport company for his personnel and freight.

Stout told Ford that since he had created the first all-metal commercial transport plane, he now had to create an organized market. Stout had always believed that there was more financial gain in the transportation of merchandise and passengers on a scheduled airline than from the manufacture and sale of aircraft. Creating the market, Stout reasoned, would directly promote his planes. Aviation had made little progress in the five years since the war ended. There were still no organized or profitable airlines in the

United States. Commercial aviation consisted largely of fixed-base operators without much capital for the purchase of a plane as expensive as Stout's $20,000 all-metal plane.

Stout concentrated on aircraft design because he knew that no airplane yet existed that could support itself in the air, financially. (Some hardly supported themselves in the air, period.) Stout, a bit naive, said, "The technical side of the work is complete. Today quantity production in aircraft is a long way off, but the day of money-making on a real scale through the operation of aircraft is very close at hand."[11] Stout's comment was optimistic to say the least. No one flew for pleasure, and few people flew on business trips, especially since there were virtually no airlines. At this point, the world still had a lot to learn about aviation. Stout went on to say, "We are not planning to sell but to operate everything we build as a measure of safety and profit where we are sure the system is safe. The final plan will include a network of airlines, and the group behind this is already taking shape, using our planes." (He was referring to Ford and his air transport service.) Stout's statement is at best a benign oversell. "With the control in the hands of a holding company like the Bell Telephone or Edison companies, we are sure to make great strides."[12] Stout was campaigning for what later surfaced as an interlocking directorate. He was astute enough to see that a unified corporate effort would result in more capital and faster advancement in the aeronautical sciences. What he did not see was the downside: price fixing and antitrust violations.

2-3 A popular plane in its day was this Travelair shown here being cranked up for a mail run.

The Air Sedan

Ford Airport

During the summer of 1924, Stout followed up his earlier discussion with Edsel Ford. In a letter he wrote to Edsel he said, "Three things are necessary to ensure the success of aviation in Detroit; a landing field where commercial development can go on without interference, and depending on government contracts; a factory on the landing field; and the increased development of reliable engines for commercial use."

Edsel gave the letter to his father, and Henry Ford told Bill Mayo to bring Stout to Dearborn to survey the land for an airfield. When Mayo and Stout narrowed the choices down to three, they went to Henry Ford. After examining the choices, Ford turned to Stout and said, "Which one do you think will make the best field?" Stout replied that the Oakwood Boulevard site would be the best but that it would take the most money to build. Ford replied, "That doesn't make any difference. If that will make the best field, that is the one we will build."[13]

2-4 The world suddenly became a little smaller when in 1924 the Douglas World Cruisers became the first planes to fly around the world. Here two of the four Douglas World Cruisers are seen at the Santa Monica Airport shortly before the flight.

Within 24 hours, Ford had moved in 38 tractors, and hundreds of men. They leveled, graded, and cleared 260 acres for the airfield, a modern factory, and a hangar.

In the *Ford News* on July 15, 1924, an announcement appeared that would have broad consequences for the world. "To encourage aircraft development the Ford interests will build a modern factory building devoted to research in aviation. The building will be used by the Stout Metal Airplane Company, and the Aircraft Development Corporation." Within six months, Stout would have his landing field with two concrete runways (the first in the country), one 3,600 feet long and the other 3,400 feet long.

Some historians credit Ford's most important contribution to aviation as the building of concrete runways at Ford Airport. Concrete runways offered a smoother surface to land, provided better drainage, and made snow removal easier. The hard runway revolutionized the airplane's ground handling also. Before the advent of concrete runways, an airplane's only braking system was the tire friction and a tail skid dragging through the landing strip's rough turf. Without brakes, taxiing was tricky and sometimes dangerous, often requiring people to hang on to each wing tip to guide the airplane. A concrete runway made both a tail wheel and main wheel brakes a necessity.

By the end of the 1920s, every progressive city in America had an airport, and many had gone to the expense of building concrete runways. The first concrete runway did not appear in Europe until the Bromma-Stockholm Airport opened in 1936.

Ford makes it official
Henry Ford, in his typical low profile, announced his entry into aeronautics. "The interest in aviation is largely Edsel's idea and he deserves the credit. Airplanes belong to another generation but I shall do everything possible in their development. The first thing to do is to make aerial navigation foolproof. Today it is 90 percent luck and 10 percent science. What the Ford Motor Company means to do is prove whether commercial flying can be done safely and profitably."[14]

On October 5, 1924, Stout and Ford had convincing proof of the value of all-metal construction. The *Maiden Detroit* was returning from the air races in Dayton, Ohio, loaded with passengers. Aboard was Professor Edward P. Warner, of M.I.T. Warner was investigating new types of aircraft for the U.S. Air Mail Service. A gas line broke, and a fire broke out in the engine. Eddie Stinson, the pilot, cut the fuel switch and put the fire out by side-slipping the airplane. He then attempted to make an emergency landing. The plane landed in a plowed field, crashed through a fence, down a ditch, across a road, through another ditch, and through a wire fence. His forward

2-5 One of the two Douglas World Cruisers left in the world. This one is in the Smithsonian Institution in Washington, D.C., the other is in the Museum of Flying in Santa Monica.

motion stopped in a farmer's backyard. The heavily damaged metal plane protected the passengers, and none suffered more than minor bruises. According to the popular story, the only fatality was a farmer's chicken. Warner was sure the all-metal construction saved him and the passengers.

Warner walked away from the wreck, and recommended that the Post Office Department buy the Stout plane as a heavy duty addition to their rickety fleet of mailplanes. Stout had sold his first plane, but ironically it took a crash landing to make his final selling point. The damaged Air Transport went back to the factory for repair, and Stout now had a problem. He didn't have enough funds to build another plane for the post office. His funds were already committed to the other aircraft Henry Ford had ordered. Stout then appealed to Henry Ford. The elder Ford, who relied heavily on the technical advice of William Mayo, asked his chief engineer, "What do you think, Bill?"

Mayo replied that he had given Stout $5,000 of his own money (at Edsel's insistence). That was good enough for Henry. Stout got the green light and additional funds to repair the plane. In December, Stout delivered the repaired *Maiden Detroit* to the post office.

On January 15, 1925, the new Ford Airport opened. Aviation experts hailed it as the finest in the country. Clarence Chamberlain, a world-famous transatlantic flyer, ranked it second only to Templehof Field, in Berlin. Stout shifted his manufacturing operations to the new factory. Jigs and fixtures were soon in place for the production of additional Stout all-metal air transports. For the next five years, Ford Airport was a center of aviation activity in the United States. Today it is the company's Dearborn test track.

A milestone in aviation

Monday, April 13, 1925, marked a significant event for the Stout Metal Airplane Company. On the ramp at the Ford Airport was Stout's second model of the Air Transport series (the first completed in the new factory), the *Maiden Dearborn I* (Stout repeated his joke with this plane, too, made in Dearborn). It took off loaded with 1,300 pounds of auto parts and mail for Ford's plant. The plane landed on time, and the world's first regularly scheduled airline, devoted solely to the business of one company was a reality.

Ford called his line the Ford Air Transport Service. Now the Fords had an

2-6 Another typical mail plane of the era, the Pitcairn Mailwing.

airline, but they also had the problem of promoting public confidence in air transportation. Both father and son believed that commercial aviation should start small with an experimental operation that would give prospective airline operators some measure of costs and potential profits. The Fords, along with Stout, believed that safety was the first consideration. The effects of passenger fatalities in the formative years of air transport would be disastrous for the industry. The Fords decided that all knowledge gained in the operation of their airline, including costs and other financial data, would be shared freely with the aviation industry.

At first, Ford's air transport service schedule called for one flight every other day in each direction. Westbound, the plane left Detroit at 3:15 p.m. and arrived in Chicago at 5:00 p.m. Eastbound, it left Chicago at 8:00 a.m. and arrived in Detroit at 11:40 a.m. Henry Ford himself kept a close watch on the timetable.[15] Ford added the *Maiden Dearborn II* to the Chicago route on April 27 and inaugurated a new route to Cleveland on July 1, 1925. A Stout-built Air Transport left Detroit at 10:40 a.m. daily, arriving in Cleveland at 12:15 p.m. The schedule called for departing Cleveland at 2:30 p.m. and arriving in Detroit at 4:05 p.m.[16] The planes shuttled automotive parts, mail, and personnel between the cities. They set an enviable on-time record, one that made the railroads take notice and wonder if there just might be something to this phenomenon of air transport.

Once Edsel was 20 minutes late for the plane to Chicago. When he got to the airport, the plane had left without him. "This is an airline" his father admonished. "We keep schedules just like the railroads."

Stout's design was sturdy, and few changes were necessary over the first year of operation. Despite the ill-tempered Liberty engine, Ford's small fleet continued to make three trips a day on the Chicago route, and two a day on the Cleveland run. Over the year, they averaged more than one million pounds of freight.[17]

To add professionalism to his airline, Ford insisted his pilots wear blue and gold uniforms. He also decided for safety reasons to put a "Flight Escort" aboard each plane.

The Flight Escorts were men assigned on each flight to make passengers comfortable, explain the functions of the plane's equipment, and point out landmarks along the route.

Everyone familiar with aviation recognized Ford's airline as an experiment. It could not claim supply and demand, for none existed. The implications were significant. The New York Tribune said, "Ford has started something and the possibilities are that he will go on to finish." Henry Ford remarked, "There is no doubt that commercial aviation can be successful. We are going to see that it is a success."

Investors who were long convinced that aviation was a hazardous investment were stimulated by the Ford example. The *New York Times* agreed. "Once manufacturers can prove that airplanes can operate as safely as railroad trains, there will soon be passenger, freight and mail airlines throughout the United States."

On the airline's first anniversary, Stout said, "We've flown the entire year without a single injury and with remarkable freedom from mechanical

Harry Gann, McDonnell Douglas

2-7, 2-8 While William Stout was building his first all-metal airplane, Donald Douglas was building his first plane, an all-wood Cloudster.

The Air Sedan 35

trouble. We've made over 1,000 trips covering a distance equal to ten times around the world at speeds close to one hundred miles an hour. We're doing this without any fanfare or publicity. We've tried to make this venture just as everyday and routine as the rest of the business."[18] Ford's airline was setting the standards and proving the value of scheduled airline operation. On May 18, 1926, right after the anniversary celebration, *Maiden Dearborn I* crashed in the first fatal accident incurred by a privately operated airline.[19]

By the second anniversary of Ford's airline, Buffalo and other cities were on the Ford route. The company reported more than three million pounds of freight and mail carried safely. Ford had proven, through a practical demonstration, the reliability of air transportation.[20]

Ford's airline and its results showed the world the potential for commercial aviation. Within a year, a growing network of the airways would begin spreading across the country.

3

The Kelly Bill

The Stout-Ford experimental operations did not start out confined to airplane design alone but were to cover the entire field of aviation, particularly the problem of flight safety. The experiments embraced every phase known to aeronautics. The factory was the first experiment; the airport the second; the inter-plant air transport line the third; and the larger airplane plant the fourth. Ford's entire aeronautical program was to be a long-term experiment. Ford was not satisfied with the safety of airplanes but felt aviation could develop into a safe and profitable industry. "I feel toward aviation as I did toward the motorcar 30 years ago," Ford said. "I could feel the interest of the people then. There is the same interest in aviation now, although not yet as high. Perhaps that is because the great masses of the people do not need aviation now, as they needed automobiles then."

The government by this time also recognized a need to promote commercial aviation. The Kelly Bill, passed on February 2, 1925, was the incentive commercial aviation needed. It allowed private contractors to bid for the right to carry airmail over routes already established by the Post Office Department. The congressional legislation said any company operating an air service could bid for a route. The winning company would receive four-fifths of the airmail revenue. This subsidy, law makers believed, would allow a commercial airline to operate at a profit while developing experience, capital, and newer equipment for a profitable passenger and freight service.

With the passage of the Kelly Bill, Ford's immediate outlook shifted to the equipment needs of airline companies forming around the country. He saw it as an opportunity to sell more planes. Everyone in aviation anticipated profits from flying mail under the Kelly Bill, and these new air transport companies needed equipment designed to meet their needs. In the summer of

1925, the only organization that had any long-term flying experience on regularly scheduled routes of any distance was the United States Post Office. They inaugurated Air Mail Service in 1918, and by July 1924 were flying the 2,612 mile New York-to-San Francisco route.[1] There were other commercial routes established before Ford, but they were small and had limited experience or had gone out of business. At the time there was only one true passenger carrying airline in the country, Pacific Marine Airways, flying a route between San Pedro, California, and Avalon, on the offshore island of Catalina. Aeromarine Airways flew between Miami and Nassau, Bahamas, in 1922, and Ryan Airlines flew from San Diego to Los Angeles, beginning on March 1, 1925 (Fig. 3-1).

3-1 This Aeromarine Airways flying boat was made of wood and fabric and in 1922 carried passengers between Miami and Bimini-Nassau the closest "wet" spot during the country's Prohibition period. Note the open cockpit in the front.

There was also the Boeing-sponsored Seattle-Victoria, B.C., mail route started in October 1920; the Gulf Coast Airline, operating the New Orleans-Pilottown route established in April 1923; and the Fairbanks-McGrath Alaskan route. With the exception of Ryan, these were not regular operators on domestic routes. They handled late mail to and from steamers, and their operations depended on the schedule of ships. The major aircraft manufacturers, like Boeing, Douglas, and Martin, until then, were concentrating on military planes and did not focus on passenger planes per se (Figs. 3-2, 3-3, 3-4).

38 The Fabulous Ford Tri-Motors

3-2 William Boeing's first commercial plane, the B-1. It flew up and down the coast between Seattle and Vancouver for almost 10 years.

3-3, 3-4 The Douglas Mailplane. Fifty-nine were built, and most went to the Post Office.

Although Ford's airline was not the only private operation, he alone focused public attention on the possibilities of air transportation. Ford published the costs for gas, oil, mechanical labor, plane and motor maintenance, salaries, depreciation, and insurance. The emerging airlines and the public evaluated the statistics on flights, loads, and completion of schedules and began to see the economic potential in the Kelly Bill.

If anyone knew the requirements for a profitable commercial operation, it would be the Post Office Air Mail Service and the men charged with the task of supervising Ford's fledgling airline. William Mayo knew that the needs of the emerging commercial airlines could best be studied in the light of practical operating experience. Mayo and Stout went to Washington in July 1925 to discuss the equipment question with Colonel Paul Henderson, director of the U. S. Air Mail Service. The discussion focused on the performance of the Stout Air Transport, sold to the Post Office. Henderson said Stout's design was unpopular with his airmail pilots. They considered it too large, too slow, and badly underpowered, especially in bad weather. The rebuilt DH-4s maneuvered better than the Air Transport, and the DH-4s had surplus power that contributed to the pilot's feeling of safety. The airmail pilots strongly resisted the change to the Stout plane or plans to buy more of them.

The Kelly Bill encouraged the operators to load up each plane with as much mail as possible. It took a full plane to yield a profit for pilots driving slow, grossly underpowered, and inefficient airplanes. The problem of overloading these planes had led to dozens of accidents and fatalities.

Henderson also believed the dangers of night flying, particularly in mountainous areas, was another reason for developing a new design. He felt the best design for this type of hazardous flying would be a trimotor aircraft. Both Mayo and Stout agreed, for they had already discussed the direction of future designs. They felt that a trimotor offered a margin of safety beyond that of any other type. Many aviation engineers felt too that no two-engine plane, fully loaded, could fly or land safely with one engine alone. A trimotor was the next logical step in design.

The dangers of flying

In 1925, flying was risky business. The carnage resulting from half-trained pilots and poor equipment was unprecedented. Dr. Jerome Hunsaker, a former MIT professor of aeronautical engineering, studied the problem and concluded that "gypsy" pilots were killing themselves at a rate of one fatality every 13,500 miles.[2] Flying's reputation continued to grow worse with each crash, and pictures of a grotesquely twisted wreckage always lingered in the public's mind. This was the major obstacle Ford and others had to overcome.

Mechanical failure was the major reason for accidents. The government had clamped down on the Air Mail Service, and by 1925 there was only one accident per 14,400 miles flown, compared to one every 2,180 miles in 1921, a vast improvement, but not good enough. The problem was the engines. The unreliable liquid-cooled, single-engine airplane was quickly pushed aside when the radial-engined trimotor came along. The water-cooled Liberty engines, for example, leaked like sieves and burned oil. The engines needed a complete overhaul every 50 hours, too. If the 50-hour mark passed without an overhaul, the engine would usually seize up, taking the plane and its luckless pilot to their doom. The early radial engines flew 150 hours without an overhaul, and by 1929, this figure rose to 300 hours.[3]

The airline and mail carriers reasoned that if a single-engine airplane had engine failure—the greatest single source of mechanically caused forced landings—twin-engine biplanes might have to do the same if one engine failed. One engine could not sustain the plane in level flight with its high drag characteristics and cargo or passenger weight. The statistical probabilities of two engines failing in flight were so remote that the trimotor offered twin-engine backup if one engine failed. Hunsaker estimated the chance of engine failure in a single-engine plane was about one in 20,000 miles. The possibilities of a forced landing in a trimotor due to two engines failing was one in 333,200 miles.[4] This was a vast statistical improvement, but not good enough for those who were the statistics.

The unfortunate airmail pilots also helped cast a negative shadow over flying. Thirty one of the first 40 airmail pilots hired by the post office died, mostly on the "graveyard run" from New Jersey to Chicago, over the fog shrouded Allegheny Mountains.[5] Conditions were primitive, without radio beams, beacon lights, weather stations, or navigation instruments. Navigation at the time was by contact, keeping in visual touch with the ground to stay on course. In fog or heavy clouds, often the pilot could only judge the attitude of his plane by the feel in the seat of his pants, a small piece of cloth tied to the center of the wing, or the sound of the wind over the wings and bracing wire. The term "blind flying" was truly descriptive. With most of the United States still rural and few landmarks to guide a pilot, the "iron compass" (railroad tracks) became the only reliable guide for a pilot to fly from one town to the next. The mail pilots compensated for the lack of technology by compiling their *Book of Directions* for the major routes. This was an unofficial compendium of railroad routes, highway junctions, golf courses, and landmarks—good, of course, only in daylight or on a cloudless moonlit night.

The United States government recognized the inadequate facilities and began sinking huge sums of money into the airmail routes. When they turned the mail over to private contractors, the airmail routes were safer. Between February 2, 1926, and April 22, 1927, for example, airmail pilots logged over 3 million miles without a fatality.[6]

In the almost 10 years the U.S. Post Office Air Mail Service lasted, 43 men lost their lives in accidents. An important legacy left by the Post Office flyers, and one that Ford learned from quickly, was the development of regular maintenance schedules that saved thousands of dollars and kept the planes ready to fly and earn money.

Stout-Ford merger

Stout had proven the validity of his ideas, and Henry Ford became a frequent visitor at the factory. One day they had a conversation on the future of aviation. Ford asked many questions about the possibilities of expanding his airline, and when Stout had answered all his questions, Ford said, "Stout, I must tell you, I'm not surprised at your deep commitment to aviation, but it looks to me as if somebody has to put some serious money behind it to make it into an industry." Ford paused and waited for Stout to respond. Stout did not have to; he knew where Ford was going with the conversation. Then Ford smiled. "I don't know why the Ford Motor Company shouldn't do just that. Talk to Mayo about us taking over this manufacturing."[7] Stout was proud of himself. He had played Henry Ford like a fine violin, and Ford's words were like music to his ears.

Henry Ford saw economic advantages to buying Stout's operation, and Stout saw the same economics working for him under the umbrella of a large corporation. Stout and Mayo (representing Henry Ford) held a long series of merger negotiations. Ford agreed to buy Stout's company stock at the rate of two dollars for every dollar invested, if Stout could convince his stockholders that they should reinvest in a new company being formed to organize an air transportation system. The *Detroit Times* looked at the proposed merger with hope for the future. "Mr. Ford's son, in many ways, is like his father," it said. "He also has the blessing of youth, and a commitment to aviation. If the Fords, father and son, undertake to build all-metal airplanes in quantity production, one great problem will be solved." Edsel Ford believed that the national defense and the country's commercial development had vital interests in the development of the airplane. Both father and son saw the enormous potential in the commercial aviation industry.

Stout knew his stockholders were not interested in the small profit from the Ford buy out, although the two-for-one stock trade was not a bad deal. They were all men of considerable wealth, and they might still sell if it meant furthering aviation's future capital gains.

"I have to break my promise to you," Stout wrote to his stockholders. "I said originally you would not see your money again, but now I have to tell you it will be doubled. But it will depend on everyone selling out."

"So, let me get for you $2,000 for every $1,000 you invested, and let me use the money to start a passenger airline out of Ford Airport." Stout went on, "It will be the first of its kind in the world, and without increasing your original investment, we will be giving Detroit another boost."[8]

Stout, the ultimate salesman, easily convinced his stockholders that there was greater opportunity in running an airline than in manufacturing. On July 24, 1925, the last checks went out to the Stout Metal Airplane Company stockholders. On Friday, July 31, 1925, Henry Ford's 62nd birthday, the Stout Metal Airplane Company shut down. It reopened the following Monday as the "Stout Metal Airplane Division of the Ford Motor Company."[9]

Quick to take advantage of opportunity, Stout called a meeting of his stockholders and new Board of Directors. The new directors included Stout's heavy investors, Henry Ford, Edsel as President, William Mayo as vice president, and several active members of Stout's former company, Fred Fisher, C. F. Kettering, W. O. Briggs, George Holley, and Sidney Waldon. Stout called the new board "probably the greatest single step toward making Detroit the world's aviation center."

Typical of the Ford organization of that day, little time was wasted on the niceties of administration. The actual transfer of control took place with so little friction that Stout wrote to Edsel Ford thanking him, "The way the smooth transition has been handled with us, by yourself and by Mr. Mayo, has been so exceptionally fair in every way."

Ford agreed to the merger because he regarded the Stout Metal Airplane Company as the most sophisticated airplane company in America and because it appealed to him as the logical, dependable, and safe company making airplanes.[10]

The short-lived 2-AT

By December 1925, Stout had manufactured 11 single-engine 2-ATs. The total number of 2-ATs manufactured is not certain, but at least 11 were manufactured. Its importance in the development of the all-metal transport flying freight and mail is clear. Ford bought five 2-ATs (*Maiden Dearborn I through V*) and used them for the Stout operation. Florida Airways operated four before going out of business in December 1925. (Florida Airways lasted only four months, selling its routes to Pitcairn Airways, Inc.)[11] (Fig. 3-5). They would be the last of the single-engine airplanes manufactured by Stout.

At least one was later temporarily fitted with a Pratt & Whitney radial Wasp engine.[12] Two 2-ATs were flying as late as September 1928, but the Air Commerce Act had put safety restrictions on pilots and airplanes and created airworthiness certificates. The licenses of 2-AT-6 and 2-AT-7 were canceled in 1928 because of structural weaknesses.

3-5 Two of the Stout single-engine 2-ATs sold to Florida Airways.

A mail route in Ford's future

On September 8, 1925, the Postmaster General, Harry S. New, went to Detroit. The visit was ostensibly related to surface mail problems, but the Postmaster coincidentally paid a visit to Henry Ford.

The Post Office Department had advertised for bids on eight proposed airmail routes that were to operate as feeder lines to the government-operated transcontinental mail line. It also had several additional routes over which it was hoping to extend the airmail system. One of these was between Detroit and Chicago, and another was between Detroit and Cleveland, both, coincidentally, routes over which Ford already operated his inter-plant airline.

The Postmaster wanted Ford to bid for the Detroit-Chicago and Detroit-Cleveland mail routes that were to become the first contract airmail routes. It was logical, said New, that Ford, who had the experience and the equipment, would be the first one to do the job. After all, the first airmail contract routes must not fail. That would be a serious defeat for the administration. Ford and New agreed on what a reasonable bid would be. Following the conference, the Ford Motor Company announced that it would enter a formal bid for the mail contracts.

A year later, the Postmaster General revealed just how far Henry Ford had gone to further aviation. "Quite a long time before this contract service, Mr. Ford offered to carry the mails without compensation over these routes. This way, the department might gather sufficient data to base its schedule and rates. We should not lose sight that the Ford Motor Company furnished transportation for its own products over these routes without which the airmail would have to be carried as a loss."[13]

There was no question that the private Ford airline was an unqualified success. As of June 2, 1925, air express from Detroit to Chicago cost Ford 8.5 cents a pound. The company dispatched an average of 32,000 letters daily, weighing 800 pounds. Ford statistics showed a savings in postage of $1,250 per day.[14]

The Ford reliability tours

The Kelly Bill gave a needed shot in the arm to commercial aviation, but it did not win any converts to the notion that aviation was safe and reliable.

In a move to provide the necessary public exposure for the 2-AT, Henry Ford created the National Air Reliability Tour. It was similar to the automobile tours of the previous generation. The planes would fly from city to city, starting and finishing at the Ford Airport, in Dearborn. A trophy and prize money would go to the winner.

To a public familiar with the accident toll of gypsy pilots and Post Office fliers, safety and reliability in air transportation were little more than words. The Fords knew that commercial aviation would not succeed until the stigma of carelessness and danger were gone from the public's perception. The negative feelings about aviation were so deep in Detroit that by 1925, the Detroit Aviation Society decided not to make its usual bid for the Annual Pulitzer Airplane Speed Classic. Instead they suggested a national air meet at the Ford Airport, emphasizing reliability and safety, rather than speed and daring. The Fords agreed to sponsor the program, and during the Summer of 1925, they began making plans for the first meet. The only requirement to enter was the ability of the airplane to carry a commercial payload at a top speed of at least 80 miles an hour.

Under the rules, each plane would be judged on its performance of a fixed schedule over a series of designated segments from Detroit to Chicago, St. Louis, Indianapolis, Dayton, Cleveland, and so on back to Detroit, a total distance of 1,900 miles. Any plane completing the circuit was a "winner" and each winner would have its name inscribed on the Tiffany trophy donated by Edsel Ford.

The event, open to all aircraft manufacturers, attracted the field's most renowned person, Anthony (Tony) Fokker. Fokker, famous for the Fokker DR-1 tri-wing war plane of the Red Baron, saw a chance to steal the limelight from potential competitors. It is not clear why Fokker took a very unusual step, but the results were to become history. He cabled his airplane factory in Amsterdam and advised them to take his newest transport model, a single-engine F-VII, and convert it to a trimotor (Fig. 3-6). There is speculation that he had seen drawings of Stout's first trimotor. When the plane arrived, Fokker entered the First Ford Reliability Tour.

When Edsel Ford dropped the starter's flag on September 28, 1925, 17 entrants raced across the field at one-minute intervals and disappeared over the horizon on the way to Fort Wayne, Indiana, about 135 miles away. Many

3-6 The Fokker F-VII that Tony Fokker flew to victory in the First Ford Reliability Tour. Immediately after the tour, Fokker flew the plane to Miami, Florida, where this photograph was taken.

types of airplanes participated, among them Stout's 2-AT (Fig. 3-7), a Travelair, a Waco, a Swallow, a Fokker, and a Curtiss. Five days later, a crowd gathered to watch the 11 "winners" land at the Ford Airport. The nation cheered the results of the Tour, and the participants said it was an example of the outstanding reliability of commercial air transportation. No one pointed out that five days was a comparatively easy schedule to cover the 1,900 miles. Also overlooked was the fact that only 64 percent of the planes managed to complete a flight that was about half the distance of a flight from New York to San Francisco.

Fokker's trimotor dominated the tour. It stopped in 13 cities, and with the help of another pilot, managed to stay out in front, all the way. Fokker came in first, but surprisingly just three minutes behind was Ford's entry, his single-engine Air Transport.

Fokker never flew in another Ford Reliability Tour, for two reasons. Ford's single-engine 2-AT gave him a close race, and the first tour did something he hadn't expected. Both Henry and Edsel showed a keen interest in Fokker's plane and the Fords bought it for Admiral Byrd's 1925-1926 North Pole Expedition. In honor of Edsel Ford's wife, Byrd named the plane the *Josephine Ford*. The publicity attendant with this single airplane was enough to launch Fokker's trimotors on a successful career in America.

The Reliability Tour captured the imagination of the public, and the committee immediately drew up plans for another tour, increasing the distance and participants, for the following year. The committee also devised an elaborate point system that would permit the choice of a specific airplane as a winner. Officials would revise the system in the years ahead to keep pace with the constant improvements in airplane design.

The Ford Reliability Tours, as they were commonly known, became a

3-7 The Stout 2-AT, *Maiden Dearborn II*, at the Ford Reliability Tour.

yearly feature for the next six years, competing in popularity with the National Air Races. The circuit constantly increased until the sixth tour in 1930, when 35 airplanes flew more than 4,000 miles. They visited 29 cities, including such widely scattered points as Moose Jaw, Saskatchewan, and Enid, Oklahoma. The seventh and last tour in 1931 covered 5,000 miles, winding through the central part of the United States, dropping as far south as Corpus Christi, Texas, and as far north as Windsor, Ontario.

Reliability tours—a significant boost for aviation

In 1931, the Tours ended. The Tiffany trophy, won in 1930 and 1931 by Harry Russell in a Ford Tri-Motor, became a permanent part of the Ford collection.

By 1931, scheduled air transportation was also an accepted fact. The reliability tours went a long way to proving that commercial aviation was a permanent fixture in the United States. The tours also encouraged people to build airports in progressive communities throughout the country.

The Tours had more than served their purpose. Manufacturers had an opportunity to display their most recent products throughout the United States and Canada. Many cities built airports to be eligible as a stopping point for the Tour and thus helped expand aviation. The rules governing entry and the point-scoring system were directly responsible for improvements in aviation and safety. Engines of higher horsepower, cabin soundproofing, and improved safety devices like the airplane brake were some by-products. Other offshoots of the competitions were higher top speeds (from 115 mph to 150 mph) and airplanes with increased safety and load-carrying ability. Attention focused on American pilots and aircraft and public education. Like the Glidden Tours in the early years of the automobile, the reliability tours brought practical aviation into close contact with the people of America. The reliability tours were just the beginning. The aviation growth that followed the first tour in 1925 saw a series of extraordinary successes, juxtaposed with a series of dismal failures and spectacular disasters. The latter part of the 1920s would be an exciting period of growth for American commercial aviation.

Right after the First Ford Reliability Tour, another significant event took place. On October 8, 1925, Leroy Manning, a Ford pilot, took off from Dearborn, Michigan, for Mineola, New York, with the first airplane sold by the Ford Motor Company to a private company. The plane was going to the John Wanamaker Company of New York. Just 20 years before, John Wanamaker had become one of Ford's first automobile customers. Now he was buying his first airplane. This would open yet another aviation market, the executive aviation market. The Wanamaker plane was a Florida Airways' 2-AT that Ford had repurchased when Florida Airways went out of business.

First commercial airmail contract

On February 15, 1926, the Ford Airport was crowded with dignitaries anxious to be part of history. The Ford Air Transport Service was about to become the first private contractor to officially fly the U.S. Air Mail.

In a letter Ford wrote to his good friend Thomas Edison and put aboard the first flight, he said, "This is really a great step forward. The pioneering in plane building and operation is behind us now. It now remains for men of business to take hold of the opportunity."

Speaking at the inaugural flight, Assistant Postmaster General W. Irving Glover said, "Under the Act of February 2, 1925, Mr. Ford and Detroit get the honor of being the first to undertake this new type of mail transport. It is really the first step in a new epoch." After the mail went aboard the *Maiden Dearborn I*, the pilot taxied down the snow-covered field and took off for Detroit, some 252 miles away. An era had begun. By June 1926, there were 12 mail lines in operation, covering a total of 6,594 miles. Eleven of them were run by private operators.[15] Within another year, 20 airlines were flying the skies over America.[16] It was not obvious at the time, but contract airmail would be the backbone and lifeblood of American Air Transport until the arrival of the Douglas DC-3.

By the end of 1926, more than $5 million had been invested in the contract airmail system. There were 55 stations, and 2,000 miles of lighted airways for night flying between New York and Salt Lake City, Utah.

The contract mail system also spurred manufacturers. In 1925, there were 24 aircraft and engine manufacturers, with a combined output of $12.5 million. By the end of 1926, there were 59 companies, with a total output of $24.1 million. American commercial aviation had come of age.[17]

In 1926 the president of Curtiss Aeroplane & Motor Company put Ford's contribution to commercial aviation in perspective: "Mr. Ford was the first in the air operating a line of all-metal ships from Detroit to Chicago, transporting his products and mail. He freely offered his experience in operating expenses to all other operators. His whole operation has been public spirited and generous and was the starting point of organized commercial aviation in this country."[18]

More than 100 airlines have flown the Ford Tri-Motor in North America, Central America, South America, Europe, Australia, and China.

Compared to other aircraft of the era, the Tri-Motor was a formidable giant: it was 50 feet long and had a 74-foot wingspan.

A Ford Tri-Motor in flight with a P-40 War Hawk and a Chance Vought Corsair.

A rare Ford Tri-Motor in flight with a B-17 bomber.

The metal Ford looked safe and dependable to a public accustomed to wood and fabric aircraft.

The thick wing root of the Ford is evident in this photo.

Ford designed the Tri-Motor to be simple to fly and maintain, believing simplicity ensured safety.

The fuselage of the Tri-Motor was made from duralumin, an alloy invented to overcome corrosion.

4

The birth of the trimotor

After listening to Mayo's report on his discussions with the post office, Henry Ford decided to go ahead with a trimotor design. Mayo had told Ford, "I am confident that with the Wright motors we cannot only fly on two motors, but can lift off the ground if the load is not too heavy and land even though operating on one motor." That is supreme confidence in the face of little or no empirical data. The Wright Engine Company began production of the first practical radial engine around 1924, and they offered Mayo three 200 hp engines for a trial.

With Ford's approval, Stout began the layout of his first trimotor. As a cost-cutting compromise to Henry Ford, Stout's design was little more than a revision of the Air Transport design. The major complaint from the post office against the original Air Transport was that it lacked reserve power. The three Wright engines, Stout reasoned, would solve the problem. The all-metal structure had proven safe and reliable, and the factory had the jigs and fixtures for the production of the basic plane.

Stout mimicked what Ford had done with his early "Model T" automobiles. Stout and his engineers decided to make the minimum number of changes in the Air Transport and to increase its performance without radical alterations in its size or structure.

But things at the Ford plant were about to change. An erroneous newspaper story named George Prudden as Henry Ford's personal representative at the airship *Shenandoah* crash, and Ford had Mayo fire Prudden. That left Stout without an integral part of his engineering team. Mayo asked Tom Towle, a graduate engineer from Yale, to replace Prudden and to do stress analysis and calculate load factors. One thing Towle noticed was that the Air Transport had a low-lift wing and engine mounts made of riveted sheet aluminum. The mechanics had already reported fatigue cracks in one of the Air Transport's engine mounts supporting the 420 hp Liberty engine. Towle worked alone investigating design changes to improve the Air Transport's low lift wing and calculated its performance with different wing sections. Even Ford's pilots off the record reluctantly admitted they did not like the airplane's sluggish lift or maneuverability. Towle told Stout and Shorty Schroeder, Ford's chief pilot, of the problems he noticed, but Stout did not seem concerned. He ignored Towle and did not order any changes in the design. Schroeder was not happy with Towle's observations. Schroeder and Towle had taken some aeronautical engineering courses at McCook Field, and Schroeder respected Towle's opinion. Schroeder also agreed with Towle's assessment that someone should substitute chrome molybdenium steel tubing for the aluminum engine mounts. Without Stout's approval, Towle designed and installed a steel engine mount on the 2-AT on the Dearborn-Hammond run. Stout was furious when he found out. (Towle's actions were justified a few months later when an aluminum engine mount failed and the engine nearly fell off an Air Transport during a landing.)

Work progressed rapidly, and on November 25, 1925, the completed trimotor, designated 3-AT for the third in Stout's transport series, rolled out on the airfield for tests (Fig. 4-1).

The 3-AT was nearly identical to the 2-AT from the tail to the landing gear, but that is where the resemblance ended. The 3-AT had a wider span,

4-1 The Ugly 3-AT.

accomplished by increasing the wing 10 feet in the center section. The engineers rounded the fuselage at the nose, but the center motor mount appeared misaligned on the cabin window. They also strengthened and widened the landing gear to bear the extra weight of the two outboard engines mounted in the leading edge of each wing. The three Wright J-4 radial engines gave the new plane a 180 hp increase over the single Liberty engine in the 2-AT. Stout had first tried three Liberty engines on the 3-AT. He abandoned that idea at an early stage when he saw the additional weight and maintenance problems they created.

With its stubby nose, the plane resembled a fat, ugly bird, and Henry Ford called it a monstrosity. Besides its looks, it was almost an aerodynamic disaster. There was no streamlining, and the engines mounted in each wing nearly destroyed what little lift Stout designed into the wing. The 3-AT was also handmade and did not have interchangeable parts, which made the assembly process difficult. Often parts were filed or shimmed to make them fit. The nose engine, mounted below the cabin window, looked like a last minute add-on, and the 3-AT appeared to be a straightforward blunt-nosed conversion of Stout's Air Transport. The open cockpit created severe drag and added to the lift problems. Stout said the open cockpit was to accommodate the airmail pilots who said they judged the attitude of the plane by the wind in their face. Another engineer said later that Stout wanted an easy exit for himself during flight tests.

Just before the flight tests, Stout called Tom Towle down to his office. In the presence of his business manager, Stanley Knauss, Stout fired Towle. Towle had a mechanical engineering degree, and that was enough for Stout to use him as a scapegoat. Towle was not happy with the design of the 3-AT. He made this known and suspected that Henry and Edsel would not be happy with the flight tests. Stout knew Henry Ford's aversion to college-trained engineers and took advantage of this. Stout figured if anything went wrong as Towle said it would, he could blame Towle. Stout was unprepared for Mayo's reaction. Ford's top engineering executive hired Towle back before the test flight of the 3-AT, to be assistant chief engineer of the newly created Aviation Division of Ford Motor Company. Stout, although embarrassed, pretended nothing had happened between himself and Towle.

3-AT an ugly albatross

Despite the ungainly appearance of the 3-AT and Towle's criticism, Stout still had high expectations for the plane. Those hopes quickly turned to despair when Shorty Schroeder flew the plane on its first official test flight in early November. A sudden stall during the normal landing approach brought on a jaw-jarring, 30-foot drop to the runway. The plane catapulted back into the air, and with a sudden burst of full power, Schroeder recovered and took the plane around again. He landed with difficulty and reported that the plane had a landing speed of 90 mph and a top speed of 110 mph.[1] A 20 mph reserve speed range was not enough power in an emergency. He had just proven that in a live test. Schroeder concluded correctly that the wing engines reduced the effective lift to the danger point. Once on the ground, he adamantly refused to take off again. Schroeder was so angry at Stout, he could only sputter. He had nearly lost his life in the monstrosity. The 3-AT was a full-blown failure.

The evening after the first flight of the 3-AT, Henry Ford quietly invited Schroeder to have dinner with him. Schroeder was only too happy to tell Ford about Stout's airplane. The stall and poor lift according to Schroeder, were the result of Stout's streamlined, double curvature, low-lift wing, compounded by the engines mounted in the wings.

Ford, always skeptical, had another pilot fly the plane. The second pilot had the same nearly fatal results and wrote in his log book, "It would be commercially impossible to operate this plane with any sizable payload, passengers or baggage. With a 1,500 pound payload the airplane will not fly level on two engines."

He also felt the plane was unsafe to land, even at a large well-paved field like the Ford Airport, with the three engines functioning normally unless "the payload was thrown out before hand."[2] He concluded the poor wing design required lower payloads, and takeoff and landing speeds too high for the rough fields that they called runways.

The 3-AT made at least three flights, contrary to some published reports. The third pilot report said that the plane did poorly because of the center section stall-out whenever the engines were throttled back and that it was almost impossible to land with power reduced.[3]

A mysterious fire

Over the course of the Stout-Ford relationship, personality conflicts and technical disagreements began to surface. The dismal performance of Stout's 3-AT was the last straw. For about a month, Ford withheld his decision on what to do with Stout's ugly albatross. Even before the misadventures of the 3-AT, there were strong hints that Ford wanted to push Stout into the background. Henry Ford, disgusted with the performance of the 3-AT, had also become disenchanted with the entrepreneurial and fickle Stout. Henry was not one to sit back and wait for meaningful progress from the eccentric inventor. He had a temper and was uncompromising when angered, and the performance of Stout's 3-AT angered him. He had a good mind to scrap the idea of Ford aviation, but Edsel was more progressive in his thinking.

Edsel proposed that Harold Hicks, a talented engineer who worked on the Liberty engine during World War I and later as a special engine design engineer for the Fords, be put in charge of aviation engineering and begin a new design. Henry Ford agreed on the condition that Stout have no part in the design work. Stout's new and only role was to develop and promote the "Stout Air Service," using the 2-AT Air Transports. The elder Ford remembered how Stout had sold him on the commercial aviation potential and recognized that this talent could make money for him. Stout was to have no further voice in aviation engineering at the Ford factory.

Stout had major differences with Henry Ford, but he never changed his belief that the entry of Ford into aviation was a great stimulation for the industry. Stout rated his achievement in selling aviation to Henry and Edsel Ford as of greater importance than his all-metal airplane. In all fairness to Ford, however, Stout did not "sell" Henry Ford on aviation. Ford's first interest began back when Edsel was 16 years old. Stout just showed Ford that the airplane had a practical commercial use.

Meanwhile, Henry Ford decided that if they were going ahead with a new airplane design, he and the public must never be reminded of the Stout monstrosity again.

Ford had a special group of employees others called his Back Door Gang, and they controlled much of the activity around the Ford laboratory. The group, according to many reports, was made up of men without much talent, but Ford had trusted them over the years. "They snooped for Mr. Ford," said Tom Towle years later. "The sweepers were told they were not sweepers, but detectives, and should report everything." Towle once asked a Back Door Gang member why Ford always shook hands with everyone. His reply: "So they can't shoot him!" Jimmy Smith, one of Ford's Gang, asked Tom Towle to stay behind one Saturday, after work ended. Smith asked Towle to put all Stout's engineering drawings from the factory into a pickup truck and take them to the Ford Laboratory building.[4] Towle knew better than to question those orders.

Late the next night, Sunday, January 17, 1926, the airplane factory went up in flames. The mysterious fire destroyed the Stout 3-AT trimotor and several 2-AT single-engine planes awaiting conversion to trimotors.

The official report said, "The only trace of the planes inside the factory at the time of the fire were the steel parts on the floor in their relative positions. Gasoline was the obvious propellant, and the intense heat vaporized the aluminum, leaving the floor covered with gray ashes."[5] There was very little said about the fire itself. There was no constructive purpose discussing the obvious, or angering Henry Ford. Besides the airplanes lost, about a dozen Wright Whirlwind engines and a Ford eight-cylinder experimental engine were also destroyed.

Stout was numb. Another disaster, but at least this time it wasn't pilot error or lack of funds. Stout refused to see the obvious. After the fire,

Henry Ford told Stout not to be discouraged, saying it was "the best thing that ever happened. Now we can build the kind of building and hangar we should have built in the first place."[6] Ford had another problem. His airline had become an integral part of his business and was carrying a sizable amount of freight, mail, and personnel for his company. He needed a new and bigger plane.

Still Stout would not give up his dream. He went to the abandoned end of the Ford laboratory, unrolled a spool of heavy paper, and took out a hammer, some glue, and some wooden mock-up sticks. He was going to pick up right where he had left off. Then, as if nothing had happened, he proceeded to build another mock-up of the 3-AT. Apparently, Stout had not caught Ford's hint, and he refused to admit that Henry Ford was not in the mood for any plane that even looked like the 3-AT.

After the fire, things were looking bad for those who had encouraged Henry Ford to get into the airplane business. Even Mayo, with whom Ford had a special kinship, being self-taught himself, was in a state of temporary discredit. No one at the Ford Motor Company could design and engineer a whole airplane, except Stout.

Mayo had asked Towle if he could design a three-engined airplane. Towle pointed out that Stout was still technically head of the Aviation Division and was already working on a mock-up resembling the 3-AT. Mayo promised Towle, "no interference from Mr. Stout" and gave him a free hand to design a new three-motored air transport. He also told Towle to supervise the immediate construction of a prototype. Towle began work on the project, in secret, using drafting boards in a far corner at one end of the Ford Laboratory. Mayo also gave Towle the go-ahead to hire several young engineering graduates to work with him on the next Ford plane.

Tom Towle told the story of why he eventually had to fire one of these new hires—James McDonnell. "Jim McDonnell had a habit of falling asleep on his drafting board every afternoon at 1:30. It was his lunch hour, but Mr. Ford usually came with Edsel around that time. They would sit in the front seat or on the running board of a car in sight of our drafting boards. Mr. Ford did most of the talking, and Edsel the listening. It was the daily tutoring session, and Mr. Ford worked hard on Edsel."[7] Jim McDonnell was still taking "naps" in 1967, but not on someone's drafting board. He used a couch in the McDonnell Aircraft Company's main office, where he served as its president.[8]

None of the new engineers had any practical experience in aircraft design. Only Tom Towle had experience with wing design, metal construction, and stress analysis, having worked at the Dayton-Wright Co., the Glenn L. Martin Co., and the Aeromarine Plane & Motor Co. The young men watched and took their lead from Towle.

Everyone left Stout alone. It was not good for one's job security to get too close to Ford's nemesis. Stout was blind to the real reason for his quasi-exile. Mayo fed Stout's ego by telling him he was a great promoter (which he was). He told Stout that Henry Ford wanted him to go on an extended lecture tour around the country. He said Ford wanted him to promote aviation and the idea of airlines using Stout airplanes. This delighted Stout. He was anxious to start his ego trip and the tour. Stout left, expecting that the three-engined air transport project would wait until completion of the new factory on the burned-out site. But Stout was still an engineer at heart and before he left, he sent Henry Ford some sketches of his revised trimotor. It was almost a twin of his original design. The sketches disappeared, and no one ever talked about them again.

The Ford Flivver

Around this time, work began on one of Henry Ford's pet aviation projects, the Ford Flivver. Hicks was promoted up to engine designer to work on Henry's project, and Towle, for all practical purposes, became chief engineer in charge of the trimotor design.

The Flivver was a small, single-place plane developed independent of the trimotor designs. The first Flivver, introduced on July 31, 1926, Henry Ford's birthday, had a three-cylinder French Anzani engine. The first showed promise and led to a second version developed under the guidance of Ford test pilot Harry Brooks, with Koppen and Hicks helping. This project was under the personal approval and direction of Henry Ford. Brooks, a good pilot, was the bright young son of Henry Ford's close friend.

The second Flivver had a two-cylinder engine of 36 hp developed by Ford. Brooks kept the small plane in his garage and flew to and from work at the plant. The Flivver was Ford's way of proving that eventually aviation would reach the common man just like the automobile. The Flivver was an open-cockpit, low-wing, non-metal monoplane that could reach 100 mph. In January 1928, Harry Brooks set out to prove the value and dependability of the small lightplane by flying it nonstop from Detroit, Michigan, to Miami, Florida. Bad weather forced him down at Ashville, North Carolina, but he set a record of 990 miles for aircraft of that type. In February, he started out again from Detroit. After an unscheduled landing in Titusville, Florida, Brooks took off for Miami (Fig. 4-2, Fig. 4-3). He crashed into the ocean on February 25, 1928, and was never seen again. Only the wreckage of his plane washed up on Daytona Beach. Ford then dropped his Flivver plans. Koppen went on to teach at MIT, and Hicks remained to design another ill-fated Ford plane, the 14-AT.

4-2 The Ford Flivver.

4-3 A size comparison of the Ford Flivver with the Tri-Motor. This Tri-Motor, 4-AT-17, was destroyed in a hangar fire in 1932.

The Towle Tri-Motor

On all Ford Tri-Motors there is a plate fastened to the outside announcing the Stout Metal Airplane Company (Fig. 4-4). The plate is misleading. Over 65 years since the Tin Goose rolled out of the factory, there have been contradictory accounts of who designed the airplane, although Stout never admitted being the designer and lacked direct involvement in its design or construction. One story says that a "design group," consisting of Harold Hicks, Tom Towle, John Lee, and Otto Koppen, designed the Ford Tri-Motor. Sometimes George Prudden and James McDonnell are included in this group. Towle was in charge of the design and hired Lee, Koppen, and McDonnell after he had worked out the basic design on the drawing board. They served as engineering assistants and draftsmen. Prudden left Ford in

58   The Fabulous Ford Tri-Motors

4-4 The plate on every Ford Tri-Motor that implies Stout designed the Tri-Motor.

1925 and was not there when the Tin Goose evolved. Hicks had nothing to do with the design, as he was an engine designer working on the Ford Flivver.

Stout returned from his selling trip two months later, surprised to see what had been going on in his absence. The design and prototype of Towle's new transport were well beyond Stout changing them. Stout had become a figurehead, and it finally became obvious that Ford wanted to use his imagination to sell airplanes, not design them. In Stout's absence, Towle and his team had created what appeared on the surface to be a new design. Towle's experience in designing wing curves (which Stout didn't see the value of) helped, but the lack of a nearby wind tunnel precluded any serious wind-tunnel testing of the new model. (After the Ford Tri-Motor had been flying, they finally got around to testing it in the wind tunnel. The tests confirmed what everyone already knew; the plane would fly.) Towle had also solved the problems of building a prototype.

Earlier in February 1926, Mayo had declared in an interview after the mysterious factory fire that "radical changes could be expected" in the future Ford Tri-Motor's design.

Harold Hicks, who had been busy on the Flivver, said later, "I relied on Tom Towle in the design of that ship. John Lee, a graduate from M.I.T., was in charge of the wing structure, Otto Koppen got the tail skid (and later the tail wheel), and James McDonnell did stress analysis, while other engineers and draftsmen helped in the details of design." He obviously avoided any mention of Stout. "We started from scratch to design the Ford Tri-Motor," said Hicks. In later years, Tom Towle would take the credit for designing the Ford Tri-Motor.

According to Towle, Mayo asked him to make three-view drawings of a new trimotor. Towle did produce these drawings unassisted since he was the only one at the time who could do the stress analysis of an entire plane for the conditions of flight known at the time.[9]

The new trimotor was not as radical as Mayo claimed, nor "from scratch" as Hicks claimed. Stout had solved most of the basic manufacturing problems they ran into with the original 2-AT and 3-AT. Stout had also worked out the problems of aluminum alloys and had analyzed many structural problems, and much of the fundamental knowledge was down on paper before the Towle-engineered plane made its appearance on the drawing board. The basic trimotor idea had even been Stout's, and the drawings were "saved" from the fire.

Towle did expand on the design of the basic exterior form, structural layout, and wing shape. He did the stress analysis and the calculations to decide its projected performance. Towle supervised the building of the prototype, and he participated in much of the early flight testing program. He rode in the co-pilot's seat while Shorty Schroeder did the piloting. He designed the refinements that gave the wing and fuselage their solid, characteristic appearance. He created the basis for the enlargements and refinements that came later.

In reality, the Ford Tri-Motor was an evolution from principles established by Stout, Prudden, Towle, Lee, Koppen, and the other engineers associated with the earlier designs.

Most of the defects in the original 3-AT were identified by Stout and Towle in the air trials, and these defects were the immediate concern in the new design. Towle's engineers, working with the NACA, worked out a higher lifting curve for the new plane's wing. While the wing proved adequate, the area behind the wing engine was blanked by the cylinder projection (similar to the problem on the 3-AT). This reduced the lift that was further degraded by propeller turbulence. The NACA cowling later overcame this problem but was not available at this time. The solution to the airflow problem in the in-wing mounted engines was to relocate the engines. Not surprisingly, it came from Tony Fokker's trimotor. The Fokker had engines in nacelles under the wings, and the Ford engineers accepted the design as a logical solution. Under the circumstances, given the state-of-the-art, it was a viable solution. Later, the in-wing mounted engine would be recognized for what it was—an aerodynamic breakthrough.

With these basic changes, another improved and streamlined version of the Stout Air Transport appeared. Out of the ashes had risen a new Phoenix, an impatient fledgling. This newly hatched Tin Goose was anxious to test its corrugated wings.

The highly respected aircraft manufacturer, Tony Fokker, recognized what Ford's efforts represented. Fokker said there was not enough demand for commercial aircraft to warrant a large and efficient manufacturing organization.

Ford ignored Fokker. Ford said the new factory would have, "every modern facility for the manufacture of all-metal airplanes, and be laid out to fit the Ford system of progressive production, now to be applied to airplane manufacture for the first time."[10] Ten years after he had approached Washington with the idea of mass-produced airplanes, Henry Ford finally had his airplane assembly line.

Henry and Edsel Ford went beyond what Fokker and others expected when they announced that all airplanes produced by their company would be multimotored types. This was an astounding declaration, since in the United States at the time, exclusive of military aircraft, there were only two or three planes of this type. In practice, Edsel Ford was setting the pace in airplane manufacturing as his father had in the automotive industry.

The decision to concentrate on multiengined planes was also in response to the market demand. It had the backing of the operators of the airlines. On the morning of the Ford announcement, the airline operators held a meeting in Chicago to discuss common problems of equipment and organization. From this meeting came the recommendation that if air transportation through mountainous areas was to be commercially successful, a multi-engine transport, preferably a trimotor, was essential.

5

Launching the Tin Goose

Despite the inconvenience of temporary quarters caused by the fire, work progressed swiftly on the new trimotor design. Just five months after the complete destruction of the factory, the first 4-AT stood ready on the Ford Airport ramp. With clean lines flowing back from a tapered nose, the plane was at least an aesthetic improvement over the 3-AT.

It was not surprising that there was a significant resemblance between the Fokker F-VII and the 2-AT. The 4-AT had three motors—one under each wing and one in the nose—a high cantilever wing (like the Fokker), and a corrugated metal skin (like the 2-AT). During this period, the Junkers Airplane Company of Germany was manufacturing and operating a low-wing trimotor monoplane also made from corrugated metal. This resemblance would later give Ford major legal problems.

When the Tri-Motor came along, it had no real competition. The Stinson trimotor was much smaller and did not have all-metal construction. It had a payload of only 1,000 pounds and a passenger capacity of only five, including the pilot (Fig. 5-1).

The Fokker trimotor was most like the Ford. It had the same engine configuration (Wright J-5 radials), and it also featured dual pilot controls. The Fokker trimotor, made of wood and veneer, looked very much like the Ford, but it was smaller in size.[1]

Tony Fokker claimed later that the Ford Tri-Motor was a carbon copy of his trimotor. He claimed that Ford had taken his trimotor used for Admiral

5-1 Chicago Southern Airlines Stinson Model "T" trimotor.

Byrd's North Pole Expedition and had engineers measure its contours and dimensions. He even repeated a rumor that Henry Ford had called Towle in, showed him a photograph of a Fokker trimotor, and said, "Build our new trimotor like that, but make it in metal."

The Ford Tri-Motor was not the Fokker trimotor in metal. The Ford's wing was different from the Fokker's wing. It was an adaptation from the NACA contours. Also many features introduced by Stout in the ill-fated 3-AT, features such as the thick corrugated skin, vertical fin and rudder, and aft cabin section were obvious in the 4-AT.

Aerodynamically, the 4-AT was a vast improvement over the 3-AT. The new Ford had a wing span of 74 feet and a thickness of two and a half feet at the root. This enormous wing spelled LIFT at little more than 50 mph (Fig. 5-2). Pilots and passengers later discovered the wing would maintain lift even in violent turbulence. In an emergency, the 4-AT could fly on two engines, thanks in part to the large wing. There have been times under ideal conditions when a pilot could maintain controlled flight on one engine (the nose engine, although finding a landing spot was then his immediate concern).

Compared to other planes of that era, with their struts, wires, fabric, and thin-sliced, wood-covered wings, the Ford Tri-Motor was a formidable giant. Although it had an awkward looking body, the high, cantilever wing was impressive. Beside its enormous wing span, it was almost 50 feet long.

5-2 Ford Tri-Motor wing root, two and a half feet thick, ensured strength and lift.

To a public used to open-cockpit, wood, and fabric-covered machines, the Ford looked safe and dependable.

Tom Towle had learned one important lesson from the dismal performance of the 3-AT. He designed the wing of the 4-AT to gradually "wash in" three degrees from engine mounts out to the tips. This delayed the burble point on the greatest percentage of wing area. Like the great bald eagle, the Tin Goose could hang on at high angles of attack.

An early problem that Ford tackled was weight versus stall speed. As an aircraft's weight goes up (or its cargo of airmail), the stall speed, that is, the speed at which an airplane's wing will stop generating lift, also increases. In other words, the more weight the plane is carrying, the faster it has to go to maintain level flight.

The technology that we know today—wing flaps and leading edge slats—were not yet invented in Ford's day. Airplane stall (often caused by engine failure or underpowered engines) was the leading cause of pilot fatalities during the most critical times of flight, take-off, and landing (Fig. 5-3).

In spite of the Ford being a somewhat physically tiring plane to fly, it was nearly stall-proof. Flaps, slats, and speed didn't contribute to its uncanny ability to lift more than two tons of payload out of pastures or to land safely in areas shorter than the length of a modern, small commercial jet. This was possible because of its enormous wing. There were several characteristics of the high wing that favored lift: its large span, its cantilever design (a slightly bent effect), and the internal bracing (eliminating the need for struts and

5-3 This was a Johnson Flying Service Ford Tri-Motor (N435H) that got caught in a downdraft and crashed. Notice the crumpled effect of the corrugated skin.

wires), commonplace today, but uncommon in aircraft design in 1926. This powerful wing was one reason for the Ford Tri-Motor's success.

The Ford engineers were also acutely aware of excessive weight. In their attempt to reduce the weight of the plane, they inadvertently engineered a weakness into the wing. Since the portion of the wing over and behind the cockpit was not load-bearing, they thought it would be safe to use a lighter gauge corrugated metal. This created a serious problem. After a few hours in the air, the area of thinner metal would sometimes crack from the constant vibrations of the motors. The skin would then peel back and create a baffle effect on the slipstream passing over the tail. This condition caused a violent shuddering and crabbing, with the tail trying to outrun the nose. The sheer physical strength of the pilot and co-pilot, probably a little prayer, and some luck got the pilot, passengers, and plane down in one piece. Engineers soon modified the design to prevent the metal fatigue.

Even in smooth air, the Ford was sometimes a handful. Some pilots have said it was the most difficult plane to keep in trim. The pilot who could trim the Ford into flying hands-off for even a few minutes was either very lucky or didn't have any passengers, or both. Ford pilots also had to develop a sense of anticipatory delay. After the pilot had flexed his muscle against the rudder, there was a delay in the ship's response. In a storm, this delay was exaggerated by the constantly changing attitude of the plane. That additional delay usually put plane, pilot, and passengers in a dangerous position. The efficient wing of the Ford was one factor that saved the day.

Tom Towle designed the prototype to have an open cockpit with a celluloid windshield. The feeling among pilots at the time was that a closed cockpit was not safe. The pilot could not feel the wind blowing on his face or detect the plane slipping or mushing in flight. "Since Major Schroeder and I would be wearing seat type parachutes on all the flight tests," Towle said, "it seemed logical to arrange for a quick exit should an emergency arise."[2] The first two 4-AT Tri-Motors had open cockpits, but all that followed had enclosed cockpits.

All pilots felt two engines were better than one, and three powerful engines spelled safety in capital letters. The air-cooled, nine-cylinder, Wright Whirlwind J5 radial engines pulled the Tri-Motor through the air with a cruising speed of nearly 100 mph (Figs. 5-4, 5-5).

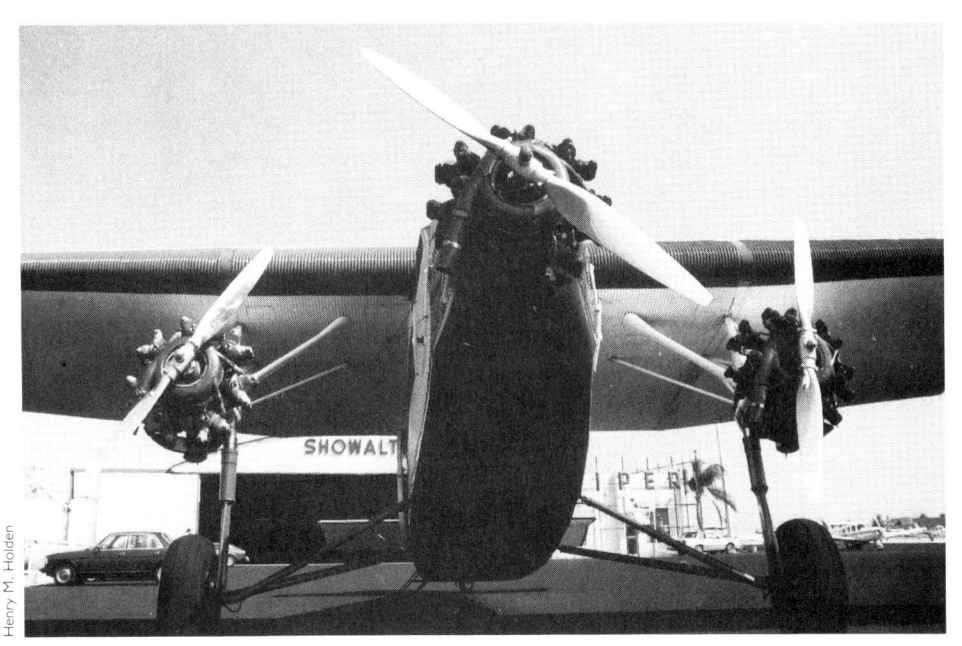

5-4 The powerful engines of the Tri-Motor.

It was the morning of June 11, 1926, and a small audience of Edsel and Henry Ford, Stout, Mayo, and other Ford executives stood watching the take-off preparations. Henry Ford was outwardly fidgety. His anxiety was showing through. He liked the lines of his new Tri-Motor—it was not ugly like the 3-AT—but he was still anxious about its performance. Towle had reassured him that it was a new design and a brand new airplane. Mayo had gone

Launching the Tin Goose 67

5-5 Another view of the powerful engines of the Ford Tri-Motor. In 1949, the J6 engines were replaced by more powerful N3N (R-760-8) engines, which N7584 has been using ever since.

over Towle's estimated and actual flight test performance figures. Mayo was optimistic, but not old Henry; he would wait and see, and then wait to hear from Schroeder.

Schroeder had more confidence in Towle's 4-AT than he had in the 3-AT. The confidence came from knowing Towle better than he knew Stout, and from having flown with Towle on some earlier test flights. Schroeder's confidence did not make Ford any less anxious. The last thing Ford wanted was a repeat of the 3-AT's performance. Seven seconds after Schroeder advanced the throttles, the Tin Goose lifted gently off the concrete runway on its official test flight. Schroeder kept the plane aloft for 55 minutes. When he landed, he declared the plane a success. America's first practical all-metal, multiengined transport was now a reality.

Ford advertising

Now that Ford had spent over one million dollars developing a reliable airplane, he had to create a market for it. For many people, flying was still a dangerous business. His air transport company was a success and he was flying his personnel between offices, but they were a captive market. They flew because the boss made them fly. The public was not yet ready to trust an airplane, not even a Ford airplane.

Since the media reported in gory detail every air crash, and was largely responsible for the public's fear of flying, Ford decided to counteract the negative press with his pro-aviation advertising. Ford financed a campaign of advertising in newspapers and magazines in an effort to persuade the public not only that flying was safe, but that a Ford Tri-Motor was the only machine to travel in, especially long distances.

Before Henry Ford allowed his name on the Tri-Motor, he wanted it safe. He was perceptive enough to know that for the public to accept commercial air travel, it would have to be perceived as safe and reliable. Ford would make his plane as safe as possible, but he recognized that air safety was a complex problem that included faulty equipment, human error, and unpredictable weather.

It was to these problems that Ford devoted untold time, money, and resources. The result was the creation of a plane whose design life was just four years, but which has lasted more than 60 years. There are 15 Ford Tri-Motors left in the world, and five are still flying. (See Ford Tri-Motor Survivors—July 1990.)

The Ford advertisements, aimed at the purchaser, the operators of the Tri-Motor, and the public, stressed safety in capital letters. Henry Ford would not sell a Tri-Motor unless the purchasers had their pilots trained by the Ford Motor Company. Ford stressed, "You don't have to crash a plane to scare your customers. Any accident, or narrow escape, raises questions." Every pilot furnished by Ford to fly the Tri-Motor had at least 1,000 hours of flying time, plus a thorough course in handling the Tri-Motor.[3]

A Ford advertisement addressed the fears of the public by saying, "All pilots of Ford planes must be men fully qualified by ability and character to assume responsibility for the property and lives in their charge. Pilots of Ford planes must have flying experience, and be examined and approved by our operations department before Ford planes, no matter who owns them, are permitted to fly with them at the controls."

Most airplanes of the 1920s did not last very long. Two years was the average life of a wood and fabric plane. So in 1929, it was surprising to read a statement by Ford in an advertisement, "No Ford plane has ever worn out in service, and we feel buyers of Ford planes run little if any danger that the planes will become obsolete before they have completed their period of usefulness, a period we know to be not less than four years."[4]

In 1932, Ford admitted that the 5-AT was designed for 2,500 flying hours. By that time, TWA had already put 3,000 hours on each of their Ford Tri-Motors and was sending them back to the factory to have the wing mailbins modified, believing the planes were good for another 2,500 hours.

Construction

The Ford philosophy was simple: The purpose of a commercial airplane was to earn money and to provide the operator with a profit. This profit came only after expenses. Therefore, Ford's goal was to keep expenses at a minimum. That is why, Ford said, the Tri-Motor monoplanes were built entirely from metal. "The metal that goes into the Ford monoplanes does not deteriorate as wood and fabric must. It doesn't rot, warp, stretch or tear. Right there is a reduction in maintenance."[5]

Another savings the company talked about was in the metal itself. Wood varied widely in strength. One piece of wood could withstand 40,000 pounds and another, apparently exactly like it, could break after 25,000. A wide safety margin was necessary. That added weight left less room for payload. Metal, on the other hand, could be accurately gauged, and excessive safety margins were not necessary.

One Ford advertisement said, "The superior rigidity and strength of metal wing construction removes the need for many trusses and braces inside, and struts and wire outside. Not only is the final weight lighter than wood and fabric construction, but parasite resistance is greatly reduced; lift and speed are both increased."[6] That was an interesting statement, and correct if compared to the drag-producing biplane design. Yet, aeronautical engineers would later discover that the corrugated skin also had high drag characteristics. While the ridges provided extra strength in one direction, they built weakness in the opposite direction, and the fuselage and wings would crumble like paper in a crash.

The Fokker trimotor (Fig. 5-6) was also popular because it was light and fast and had a smooth skin (of lacquered fabric) that created less parasitic resistance than the corrugated metal of the Ford.

The Ford planes were protected when adapted as seaplanes. In actual operation, the surfaces of the plane were exposed for up to 18 months to sea water without any deterioration. All the exposed parts of the Ford plane made of duralumin were coated with a moisture proof protection of "lionoil." These parts were not subject to wear, friction, or abrasion, and the coating was positive protection against corrosion.

The steel axles got special care to ensure they did not rust on the inside. Before assembly, a worker drilled holes at each end of the axle, and then boiled it in oil for 20 minutes. After boiling the axle, he drained the tube and

5-6 A Fokker F-VII of Pan American Airways. Photographed in Miami around 1929.

plugged the holes. He then coated the entire tube with cadmium plate, a rust resistant.

These construction techniques removed all questions about the problem of using Ford Tri-Motors as seaplanes, or as land planes in areas next to salt water. The Tri-Motor was ready for duty along the seaboard, over ocean stretches, or over land.

Molding public opinion

Henry Ford, aware of the power of public opinion, said he would do whatever was necessary to influence the acceptance of his airplane. He also would be careful about what he said. "A famous financier," he said, "president of a great railroad, in a thoughtless moment of irritation barked out, 'The public be damned.' The public heard about it, and for thirty long, hard years, gave the railroads a demonstration of the force of aroused public opinion. It shook the railroads to their deepest foundations; and nearly ruined them."

Ford was trying to mold public opinion on the side of aviation and cautioned his pilots, "The next time someone runs in front of your plane as you're taking off or landing remember, when you speak to him about it, he's part of the public. If you're asked a silly question, like, 'How do you get down if the engine stops?' remember it is prompted by an interest in aviation and requires a courteous answer. Little things to be sure, and hardly worth noting, but it's the sum of these little things plus the big ones that make public opinion. And today's public opinion is deciding the future of commercial aviation."

Ford wanted his planes to inspire public confidence and preference for air travel, so he reported on his plane's popularity. "More than sixteen-thousand men, women, and children came out to the Ford Airport at Dearborn last summer to fly," a Ford advertisement claimed. The number of people who flew that first summer may be overstated, but the Ford Tri-Motor did draw curiosity.

Comfort

Ford also recognized that safe and rapid flight were not enough to attract passengers. He saw the need for the comfort of quiet, enclosed flying. He said, "People will always use airplanes where time is vital," he said, "but compare the number of people who must hurry to the total number of people who travel. A small percentage! The big market for passenger service is the public who will fly in preference to using other means of transportation, not because emergency impels. The source of that preference must be comfort."[7]

Ford extended the philosophy of comfort beyond the airplane by equipping the airport with a comfortable waiting room, rest rooms, and ticket office. There was also a small hotel, restaurant, and ground transportation in a model-T bus. Air travelers had all the "modern conveniences to which the traveling public is accustomed," another ad said.

Ford advertising stressed heavily the comfort of the Tri-Motor. "In the Ford monoplane, comfort has been given the same attention as the structural strength and operating economy. The entire fuselage is enclosed with plenty of windows to permit good visibility and ventilation. The fuselage is below the wing, lowering the center of gravity, widening the view, and sheltering the passengers from the sun (an interesting reason for the high wing). Exhaust manifolds throw the sound away from the fuselage, and padding of the compartment further muffles sound. Conversation can be carried on with ease. Large upholstered chairs assure riding ease for twelve passengers (Fig. 5-7). There are lights for night flying and the cabin is heated. Each plane has a lavatory and a washstand, and an observation room forward."[8] (Author's note: There were very few navigational aides on the commercial routes for anyone who dared fly at night, and what little heat there was in the plane escaped through the fresh air ducts.)

The statement that conversation was easy must be put in juxtaposition with reality. D.W. Tomlinson, one-time chief pilot for TWA, flew thousands of hours in Ford Tri-Motors. "Flying in the old Fords," he said, "was almost an ordeal from the passenger standpoint. The flight was almost deafening. The old metal Ford shook so much, and the noise of the engines, well, you'd be deaf for a couple of days after you arrived at your destination. It was an uncomfortable experience. I was always surprised that people would pay money to ride in the things."[9] Tomlinson, would in later years need hearing aids for both ears, his hearing all but destroyed by the roar of the Tin Goose.

5-7 What the ad said and what passengers got were sometimes two different things. This is the interior of a Pan American Airways Tri-Motor. Note plain wicker chairs, no cushions, and no seat belts.

Simplicity

Ford designed the Tri-Motor to be simple to fly and maintain, believing simplicity ensured safety. Some people took the idea to an extreme. The *New York Times* of February 2, 1930, ran a photograph of James Terry, an inventor, making an in-flight repair. The photograph caption read, "Another step to greater safety in passenger flying." The plane appears to be flying straight and level, so it is obvious that the nature of the repair was just for publicity.

In reality, even if a repair was necessary, it is unlikely anyone would climb out and crawl upside down to the tail section and repair the problem. An in-flight repair in the tail section would have disturbed the center of gravity, resulting in excessive pitching and yawing. The mechanic would have had to make a choice, either hold on for his life, or hold on to his tools and take his chances.

Another problem introduced by simplicity was the windshield. The Ford automotive engineers designed a straight piece of glass for the Tin Lizzy. The engineers who designed the Tin Goose did the same thing. When night flying in the Ford became practical, the pilots found a confusing display of reflections off the windshield from lights on the ground. A night approach into the airport lights and the lights of the nearby city became a kaleidoscope of illusion and guess work. Pilot protest caused a change that angled the windshield and reduced the problem. Ironically, the same problem of reflected glare would arise again with the Boeing 247 and the Douglas DC-1.

The simplicity extended to the passenger cabin, too. In the first Fords there were no seat belts, only hand grips, to help stabilize the passenger. If a passenger used the lavatory, he found another simplification. The toilet consisted of an ordinary seat and cover. When someone raised the cover they got a revealing look at the passing scenery, 5,000 feet below.

Simplicity was sometimes taken to a dangerous level. The oil pressure and oil temperature gauges for the engines were outside, on the engine struts. Strange as it may seem, the pilot could see them from his seat in the cockpit, and they were illuminated for night flying (Fig. 5-8).

5-8 Engine oil and temperature and pressure gauges mounted on the engine strut were easy to see from the cockpit.

Ford continued the two-pilot idea started by Stout. The cockpit consisted of two side-by-side seats for the pilot and co-pilot. One man could fly the Tri-Motor, but having two pilots and two sets of controls was another way to give the public more confidence. The dashboard was simple and not cluttered with instruments (Ford thought of simplicity as safety) (Figs. 5-9, 5-10). The wooden steering wheel (from the Model-T) was also simple. The rudder pedals responded left for left, and right for right, and the wheel brakes were hydraulic, but not foot brakes. In the Tin Goose, to slow down or stop, the pilot rotated a standard "gear shift lever" brake mounted on the floor between the two pilots. Straight back for both brakes, back and to the left for the left wheel brake, and so forth. The braking system was a challenge during a crosswind landing. Many cockpit switches also came from the *Tin Lizzy* automobile.

Some controls were not so simple or convenient. The trim control, a small crank, was located above and behind the pilot. It moved the entire hor-

5-9 The simple Ford Tri-Motor instrument panel of N-7584—one of the last Tri-Motors still flying. Note the wheel brake lever tied back with the seat belt.

izontal stabilizer (Fig. 5-11). Since the Ford Tri-Motor did not have hydraulic assists on the surface controls, sheer human strength opposed wing and rudder loads.

An early Ford manual described the pilot's compartment. It said the large sliding windows afforded both protection and maximum visibility for the pilot. It went on to say, "The ability to fly in bad weather is largely determined by vision. If a pilot can see, he can fly. Pilots must have motorcar vision ahead in the air and on landing, especially since more planes are now taking to the air in bad weather." It also described the compartment as the strongest part of the plane and, "a much steadier place for the pilot."

5-10 The modern instrument panel of the German JU-52 trimotor, still flying.

5-11 Horizontal stabilizer on 4-AT-38. The original Ford Tri-Motor came with a tail skid instead of a wheel.

The cockpit was not spacious, but the visibility was excellent in all directions. If the pilot had to see beyond the 180 degrees the cockpit offered, instead of crabbing the tail, he just looked back through a window in the passenger cabin.

Cosmetically the Ford Tri-Motor was simple, too. The Goose's skin could be washed down with a hose and brush, and it seldom needed repair. The engine parts were easy to access. It cruised at 107 mph, held 231 gallons of 80-octane gas, and went about 570 miles on a tankful, just under 2.5 mpg. The gravity-fed fuel tanks, two large and one small reserve tank, were in the center section of the wing. "Gasoline is kept in tanks suspended in the duralumin wings, away from engines, crew and passengers. This, with the other all-metal construction, removes the fire hazard even under the most extreme conditions," said a Ford advertisement. The tanks were removable through the cabin ceiling if inspection was necessary, and there was luggage space and a mail compartment built in the wing.

Simplified part replacement was another advantage. Every part of the

Ford Tri-Motor could be replaced directly from stock with no time or labor expense in machining or adjusting the new part to fit. Again the Ford philosophy was so simple and pragmatic that it remains prominent in the industry today. Every hour of ground time saved is an hour the plane can be in the air, earning revenue.

The air-cooled Wright Whirlwind engines also added to the ease of maintenance. They did not have the complex, cumbersome, and unreliable plumbing found in the liquid-cooled engines. They were also 25 percent lighter than the equivalent liquid-cooled engine with a higher power-to-weight ratio.

First airplane assembly line

Ford's airplane assembly line was unique because it was the first. It also characterized the typical Ford production idea. The same method used to manufacture automobiles was successfully applied to airplane production.

The receiving department, at one end of the factory at Ford Airport, took in the "duralumin" and other raw materials. The materials progressed through a continuous process of molding, shaping, and forming until a finished plane emerged at the other end. Ford produced Tri-Motors valued at more than $12,000,000, and unlike many massed-produced products, the Ford Tri-Motor's evolution was constant, instead of jumping from model to model. Many changes were minor and incorporated on the production line whenever it occurred to the engineers to adopt them.

The duralumin fuselage of the Ford Tri-Motor had an "Alclad" alloy as a covering. The Alclad was an alloy invented to overcome the drawback associated with all-metal airplanes—corrosion. Pure aluminum does not corrode, and only when it is alloyed to provide strength does it become susceptible to corrosion. The Alclad alloy, sandwiched between the two surfaces of noncorroding, 99.7-percent aluminum combined the corrosion-proof quality of pure aluminum with the strength of duralumin. The Alclad alloy could be rolled into a sheet as thin as paper and still have the same proportional strength of pure aluminum on the surface. The fuselage was formed in sections and assembled with gussets and rivets, similar to a steel bridge.

Stressed skin

At the time, there was another metallurgical application describing a new technique applied to metal airplanes. When the Ford Tri-Motor was on the drawing board, technical papers were available describing a technique called "stressed skin."

German engineer and scientist Adolph Rohrbach, who experimented with Hugo Junkers' all-metal construction, calculated that eliminating the metal corrugations reduced skin friction and improved the airplane's efficiency. By 1919, he had discovered that a lightweight, smooth metal skin could bear a good deal of the load if the skin had a box-like arrangement of metal spars for interior reinforcement. It was a semi-monocoque (Pronounced mono-cock) type construction and a significant step toward eliminating the wires and braces in aircraft design. Rohrbach experimented with duralumin by positioning two pieces of this metal and welding them together, thereby increasing the strength of the metal. He called this "stressed skin," and applied this veneered metal to a series of planes he built in 1920. The four-engine E-4/20 airliner that Rohrbach designed is generally regarded as the prototype of the modern airliner. Under the terms of the Treaty of Versailles, the Allies seized and destroyed the planes in 1922. When the Allies' control on Germany's aviation industry relaxed around 1925, Rohrbach built a successful large multiengine flying boat for Lufthansa using his stressed skin.

By 1925, Rohrbach had also refined his technique using newly developed aluminum alloys. In a lecture the following year, before the Society of Automotive Engineers in Los Angeles, Rohrbach discussed the possibilities of putting all the airplane's dynamic load on a smooth "stressed skin."

5-12 The early designs of Northrop resulted in the *Gamma*, a low-wing monoplane used by TWA as a fast mailplane.

Launching the Tin Goose

5-13 The Boeing Monomail. Note the pilot sat behind the wing and the passengers were up front.

At the time, the only all-metal passenger planes in the United States were the German Junkers and the Ford 2-AT. Both had corrugated skins running span-wise, creating additional aerodynamic problems over the wood and fabric planes. The drag in the grooves was much worse than first believed. Although the corrugations stiffened the skin in one direction, they rendered the skin unsuitable for bearing loads perpendicular to the corrugations.[10]

Most Americans who listened to Rorhbach or read his papers did not experiment with this stressed-skin technique. One man, Jack Northrop, did. Northrop had been working with monocoque and stressed skin construction, first in wood, then in metal. Northrop applied this technique to the fuselage of his designs but concentrated on the wings. His design had several non-parallel aluminum spars that formed a honeycomb effect. The exceptional strength came from individual sections placed together to form a multi-cellular structure, creating a unit stronger than its parts (Fig. 5-12).

The NACA was discovering new data. Their data was coming from a series of wing and fuselage sections tested in their wind tunnel.

While Ford shared many of his early results with the aviation industry and the public, he was extremely reluctant to introduce newer technology into his airplane designs. This is shown by the variety of trimotor designs that followed. He stuck with the corrugated metal skin and high wing even when Boeing produced the stressed skin monomail in 1930 (Fig. 5-13).

6

Slow growth

There was no question that Henry Ford believed his Tri-Motor would revolutionize travel in America. To jump-start sales, the Ford Motor Company was its own first customer. Ford bought the plane to stimulate sales and to replace the 2-ATs on his air transport line. The first 4-AT, number NC1492, went into service on the Ford Detroit-Chicago freight run. The Ford Company went on to buy 17 more Tri-Motors. Maddux Airlines was a close runner-up, buying 16 Fords, but Pan American Airways flew more Fords than any other airline. They bought 10 new Fords, and 18 more over the course of six years. Pan American also had the distinction of purchasing the last Tri-Motor built, in June 1933. One Central American airline, Transportes Aereos Centro Americanos (TACA) flew 25 Fords through the 1930s and 1940s.[1]

Two straight production lines started with the capacity of two planes a month (Fig. 6-1). Ford anticipated a day when demand for the Tri-Motor would increase, and he was ready to meet the needs of the air transport companies. He was ready to produce one plane a day if necessary.

For the present, Tony Fokker had been right. A large market for multiengined planes did not exist. Many new airmail carriers, most of whom had inadequate capital, could not afford a $50,000 Tri-Motor. A large passenger demand had not developed except along the New York-Chicago route, and airmail loads and revenues were also not coming in as expected.

With production under way in 1927, Tom Towle left the Ford Motor Company to start his own aircraft company. His mentor, William B. Mayo, loaned him $250,000 and served as Chairman on the Board of Towle's new enterprise that specialized in designing, engineering, and building flying boats and amphibians.

6-1 The Ford Production Line.

By December 1927, Ford had sold 14 Tri-Motors, not a bad showing on the surface, but only five went to commercial airlines. National Air Transport bought one, Stout Air Service bought two, and the remaining two went to Maddux Airlines.[2]

Maddux Airlines was another Ford idea. Air and freight service on the West Coast was limited and unreliable, and Ford saw the potential market. Ford wanted to use his automobile dealers to sell his airplanes. One dealer was John Maddux, in Los Angeles. Ford suggested Maddux start a scheduled passenger service. Maddux did not believe the Tri-Motor was safe or could navigate over the mountains to California. When Ford heard this, he dispatched a Tri-Motor to Los Angeles. It arrived safely, and on July 21, 1926, Maddux Air Lines began flying the Los Angeles-San Diego route.[3] Maddux Air Lines grew, and during its few busy years, lost only two planes, one in a mid-air collision, and the other in a landing accident. The first Maddux crash gave the Ford Tri-Motor the dubious distinction of being the first United States airline plane to be involved in a mid-air crash.[4]

As slow as aviation growth seemed, within a year the Ford Tri-Motor penetrated every market available to airplanes: business travel, freight, private travel, and the scheduled airline market. With the Ford name appeared the colors and logos of Standard Oil, American Airways, Eastern Air Transport, Braniff Airways (Fig. 6-2), Pan American Airways, Northwest Airways (Fig. 6-3), Transcontinental Air Transport (TAT), and Pittsburgh Airways (Fig. 6-4). National Air Transport (NAT), an early subsidiary of United Air Lines, purchased the second 4-AT off the line. TAT and NAT each purchased 11 Tri-Motors.

82 The Fabulous Ford Tri-Motors

6-2 As soon as the DC-3 appeared, Braniff retired its Fords and replaced them with DC-3s.

6-3 The Ford Tri-Motor 5-AT-58, NC8419, of Northwest Airways, flew the snow routes of Chicago, St. Paul, Minneapolis, Omaha, and Winnipeg.

6-4 The Ford Tri-Motor 4-AT-54, NC9611, of Pittsburgh.

Executive Ford Tri-Motors

The Standard Oil Company of Indiana was the first large corporation to purchase a Ford Tri-Motor, and the first to use it as an executive plane. Royal Typewriter was the second, followed by Texaco and Firestone. Even the Bureau of Air Commerce, the forerunner of the FAA, bought a Ford (Fig. 6-5). Owners of Ford Tri-Motors found many uses for their planes, some unusual, like serving as ambulance planes or hauling pipe to the oil fields of Columbia.

6-5 An Army 5-AT-D model (5-AT-101) was used by the Secretary of Air Commerce, the forerunner of the FAA. This Ford came equipped with speed rings and wheel pants. Both added a few miles-per-hour to its speed.

In the late 1920s, aviation was far from a household word, and only a handful of Tri-Motors were bought new by individuals. Marcell N. Rand, an industrialist, was the first, and three other persons are listed as purchasers of the new Tri-Motors: Prince Bibesco, President of the Romanian Federation Aeronautica Internationale (5-AT-88); H.H. Timken of Canton, Ohio (5-AT-89); and the most luxurious, costing $15,000 over the f.o.b. price, went to Marshal Chang Hsueh-Ling of Peiping, China (5-AT-99).[5] It had a dining room set, an upholstered sofa and divan, wall paneling, special lighting, wallpaper, a writing desk, heat, and running water. At least two other Fords found their way to China in those early years, but they eventually faded into obscurity.

Stout Air Service

In the early 1900s, Ford had difficulty selling his "horseless carriage." People were reluctant to give up their buggies, but his persistence paid off. It was the same with his airplane, only this time people were reluctant to travel the skyways. Even without the demand, Ford continued production. He used advertising programs, public statements, and demonstration flights to lead the public into the air age. Here William Stout acted as a major force. His unique ability to dramatize a cause led many business and financial leaders to pressure airline operators to buy the new Ford planes.

Stout invested in the more promising airlines like Western Air Express and Northwest Airways, but he was not content to be just a financial investor. He wanted to be part of the action. Stout's airline offered the unique feature of carrying passengers in preference to mail or freight. On September 3, 1926, when he incorporated the Stout Air Service, he had Henry Ford's full blessing. Ford had put aside his bad feelings about the 3-AT for a business arrangement that promised a profit. The Stout Air Service became the first regularly scheduled passenger line in the United States to survive the trials and financial disappointments of the founding years. By 1929, his routes included Cleveland, Chicago, South Bend, Battle Creek, and Kalamazoo, and he could lay claim to being the oldest and best-known operating passenger line in the country.

Stout's airline prospered when others failed, through a combination largely ignored by other airlines at the time. A Detroit hotel porter agreed to sell tickets for his airline, and as Stout said, his was "the first hotel airline ticket office in the United States." The Detroit Street Railway Company provided special bus service from downtown hotels to the Ford Airport terminal, the first airline "limousine" service in the country. Ford's Airport featured the first terminal buildings specifically designed for an airline, with a waiting room, ticket office, baggage, mail room, and a pilot's room. Across the street from the airport was the Dearborn Inn, built for the convenience of the air traveler who wished to stay overnight without traveling into downtown Detroit.

The Stout Air Service became a valuable asset to Detroit and to Ford, providing a vital link to the expanding commercial airline network. By October 1, 1928, Ford planes had completed more than 5,000 trips between Detroit, Cleveland, Chicago, and Buffalo. Combined these trips totaled more than 11,365 flying hours and covered 1,006,255 miles, a distance greater than 40 times around the world. During this time, Ford planes carried 6,130,316 pounds of freight and 32,031 pounds of U.S. mail.[6] In 1929, United Aircraft and Transport's subsidiary, Boeing Air Transport, planned a passenger service from San Francisco to Chicago. With the Stout operation in the fold, the line could extend to New York. In April 1929, in a stock exchange agreement, Stout Air Service became a subsidiary of United Aircraft and Transport. This buy out gave Stout a substantial income, and with it he left the field of operating aviation. He kept his other financial investments and devoted his energies to experiments as unique in their field as the all-metal airplane was to aviation. Stout left an indelible mark on aviation; in design, manufacture, and operations. His was a genuine creative force pioneering the frontiers of aviation.

Although the Stout Air Service was one of the first to buy Ford Tri-Motors, public confidence in aviation was still not strong enough to supply either investment capital or passengers for the creation of other profitable airlines. In May 1926, the Air Commerce Act brought some regulation and control to the industry. It helped eliminate unqualified pilots, inadequate equipment, and unsafe flying conditions that contributed to the dangers of flying. By March 1928, 4,555 aircraft had been licensed or rejected for licensing.[7]

Charles Lindbergh

On May 21, 1927, America's attitude toward aviation changed almost overnight. At 10:21 p.m. Paris time, a slender young American in a silver monoplane landed at Le Bourget Field. Charles A. Lindbergh had become the first man to fly nonstop 3,610 miles across the Atlantic alone, from New York to Paris. The world went wild with excitement, and he became an instant hero. Lucky Lindy became a household word, and Americans suddenly felt involved in aviation. Congress promoted him to the rank of colonel and awarded him both the Congressional Medal of Honor and the Distinguished Flying Cross. Lindbergh used his fame to promote aviation. He took the *Spirit of St. Louis* on a nationwide tour, visiting every state in the union (Fig. 6-6).

6-6 Lindbergh's *Spirit of St. Louis* now hangs in the National Air and Space Museum.

Everywhere Lindbergh went the people were enthusiastic. Wall Street and the public who had just months before shunned aviation hurried to get on the bandwagon. Americans, mostly businessmen, all wanted to buy tickets to ride in an airplane. After all, if Lindbergh could fly all the way to Paris behind one engine and without a radio, it must be safe to fly from Detroit to Chicago in a plane with three engines, especially if it was a Ford plane.

For months the newspapers filled their pages with aviation stories. Three months after Lindbergh's flight, the *New York Times* was still running dozens of articles daily about aviation. In the August 27, 1927, issue, there were 74 different articles and features on aviation.

Lindbergh was largely responsible for the sudden attention to aviation stocks, but America was slowly recognizing that aviation was an economic force to watch. Aviation stocks had moved slowly before Lindbergh's flight, but by May 1929, the stock index had risen 1,164 points, and aviation securities were 11.2% of new issues. Organized air transport companies increased from 23 in 1927 to 97 in 1929, while operational aircraft increased from 128 to 525 in the same period.[8] America had grown wings.

Henry Ford never owned a Tri-Motor for his personal use. In fact, despite his efforts to popularize aviation, he had never even been up in an airplane. This changed when Lindbergh visited Ford Airport as part of his nationwide tour.

On August 11, 1927, Lindbergh took Henry Ford for a ride in the *Spirit of St. Louis*. Since the plane only had room for one, Lindbergh improvised and Ford sat on a wooden box during the flight. Ford returned the courtesy by letting Lindbergh fly the Ford Tri-Motor. Ford went along for the ride, and it was the only time Henry ever flew in one of his Tri-Motors.

The 5-AT Tri-Motor

As American industrial technology advanced, so did improvements to the Tri-Motor design. Engines of greater horsepower, better cabin sound proofing, and other changes that increased safety and load-carrying ability were added to Ford's Tin Goose.

At the suggestion of Charles Lindbergh, who had become technical advisor to Transcontinental Air Transport (nicknamed the Lindbergh Line), Ford developed a larger, more powerful version of the Tri-Motor, powered by three 420 hp Pratt & Whitney Wasp engines.

The 5-AT became the backbone of scheduled air transportation in the United States. Practically all the airlines at the time in the United States operated one or more Fords. (The exception was Delta Airlines.) Despite its noisy and uneconomical operation, it brought a whole new standard of dependability and service to the growing industry.

The 5-AT series was the same basic airplane as the 4-AT, but it was two feet longer and four feet had been added to the wings. The plane could carry almost twice the payload of the 4-AT.

With the 420 hp Wasp engines, the 5-AT could maintain a maximum speed of 135 mph. In test flights with a full load of 13,500 pounds and one engine dead, the speed dropped to 113.8 mph. Its range was 510 miles on 277 gallons of gas, slightly better than 1.8 mpg.

The original maximum speed of the 5-AT slowly increased. By October 1930, it was up to 152.5 mph. This added speed came from refinements of fuselage and nacelle design. On September 29, 1930, the Ford Tri-Motor set a record and became the fastest multimotored plane in the world. The single 7-AT (later modified to 5-AT-79A) won the 1930 Ford Reliability Tour, reaching 170 miles per hour.

The first 5-AT flew on July 21, 1928, and its performance was like its sister, the 4-AT, only better. The remarkably improved performance was a stroke of genius by Ford engineers. The center section of the wing attached to the fuselage and carried the outboard engines. The wing extensions were secured to the center section with six bolts. The genius lay in extending the center section four feet. This saved the expense of manufacturing two sets of wings, one for the 4-AT and one for the 5-AT.

The 5-AT came in what the Ford company described in their advertising as a "deluxe club plane, a winged yacht, beautiful as a jewel, comfortable as your club, equipped for high-speed flying without sacrificing any of the luxurious accommodations." The usual arrangement for a Club model was nine passengers and a crew of two. It had a kitchenette, folding berths, a writing desk, a bookcase, seven overstuffed chairs, a two-place divan, card tables, a refrigerator, a tiled lavatory with a toilet, running water, and towel racks.

The 5-AT progressed through various changes, none obvious until the 5-AT-D model. Tri-Motors 5-AT-4 through 5-AT-45 were B models and 5-AT-46 through 5-AT-96 were C models. There are no outward differences between the B and C models. There were also many variations within the groups themselves, illustrating Ford's assembly line technique incorporating changes when the engineers deemed them necessary. A Ford official in 1929 said, "Were it not for our system of constantly improving our product, our output would have been much larger. This improvement of design is still being made, despite a clamor for greater production. Since the tri-engined type was standardized by us there has been no product that can be termed a 'new model' or a radical departure from the several products immediately preceding it. Many new features were added gradually and at such times as their beneficial introduction occurred to the company."

The 5-AT used a rearward sloping windshield, which helped eliminate dangerous nighttime reflections and presented less head resistance.

In 1929, engineers modified a 5-AT-B completely, fairing the engines into the wings. The plane showed the expected increase in speed, but the landing speed increased, too. The engines in the wings acted as a spoiler at high angles of attack, and the lift fell off badly. Since landing speed and distance were prime factors in airline operation, the engineers went back to the original design.

With the 5-AT-D model, the first external change became obvious. The D Model was eight inches higher than the C Model, a difference that is easy to notice when the distance between the top of the cabin windows and the bottom of the wing is compared between the two models.

During the early preparation of the *Ford Story*, William T. Larkins found it impossible to find any good data on the 5-AT-D. Through careful observation, he discovered the significant change "by comparing photographs," he said, "of the C and D, and measuring the number of corrugations in the fuselage skin between the top of the windows and the forward bottom of the wing, of the 5-AT-C. From the Parts Price List it was known that the height of the glass in the cabin windows was 13 and $^{11}/_{16}$ inches. Again studying the photos, with a 7X jeweler's glass, it was shown that this figure equaled twelve corrugations. Dividing one by the other gave 1.14 inches per corrugation. This figure, times the seven, comes out 7.98 inches. Two years later, when the official figures for the 5-AT-D were finally uncovered, it was found that the difference in height between the 5-AT-C and 5-AT-D is eight inches."[9] The greater height allowed more headroom in the cabin, and allowed freer inspection of the wing center section and fuel tanks.

The person given the most credit for developing the 5-AT was William Mayo, but his position was mainly administrative. A Detroit paper discussed the part that Mayo played in the development of the 5-AT:

> The patron saint of American aviation, in certain of its phases, is William B. Mayo. It was through him, that Henry Ford became interested in and finally purchased the Stout Metal Airplane Company, putting his vast resources behind the development of practical airplanes.
>
> It was under William Mayo's inspiration, also, that the first exclusively passenger airline in America was started, and he inaugurated the world's first wholly freight-carrying airline. Ever since, these facilities have operated with an astonishing record for safety and efficiency. The freight planes have carried more than ten million pounds of Ford material.
>
> When he passes on to his reward Henry Ford will leave behind him many evidences of his genius. None will be more striking than the fact that he gave William Mayo his great opportunity.

Supply and demand

Air travel was catching on, and Ford was there to meet the demand. Ford's manufacturing activities easily kept pace with the demand. Nineteen twenty-nine was the biggest year for the manufacture and sale of Ford Tri-Motors.

Ford sold more than 50 percent of the multiengined transports sold in this country that year. During the year, Ford delivered 91 planes to customers, 35 more than in 1928. It was possible for Ford to manufacture as many as three planes a week. To do this, Ford expanded the factory's floor space 155 percent in June. Late 1929 brought another addition to the factory, now employing 1,850 men. Ford did with the airplane what he had done with the automobile.

Ford had long promised to reduce the selling price once he got mass production underway. The increased production capability enabled Ford to bring down the selling price of the Tri-Motors. He reduced 4-ATs from $50,000 to $42,000, and 5-ATs from $65,000 to $55,000.[10]

The stock market crash

October 29, 1929, called "Black Tuesday," spelled disaster for aviation stocks and sales, as it did to every other phase of American industry. The stock market and the world economy came crashing down. A wave of selling starting on October 24, 1929, and ending on October 28 resulted in a $14 billion loss on the New York Stock Exchange, and the plunge of the United States and the world into a depression that lasted years.

Tri-Motor sales reflected the depression economy. By early 1930, Tri-Motor sales dropped to 14. There was, however, a contradiction. Even with a world-wide depression, something was happening in American aviation. It was slowly and inexorably moving forward. In 1930, the dawn of a new era was beginning. The Commerce Department reported that scheduled air passenger travel, both domestic and foreign, had doubled over the previous year. Passenger volume went from 173,00 in 1929 to 417,500 in 1930, and total miles flown increased from 25,141,400 to 36,945,200. On the domestic scene, there were 47 airlines flying 165 routes. During the last half of 1930, those airlines covered more than 17,300,000 miles. There were another 27 mail lines that would accommodate passengers if they had room. Clearly, the United States was taking the lead in air commerce, but something was also happening at the Ford airplane plant.

In 1931, Ford sold only 22 Tri-Motors, and by 1932, he saw the handwriting on the wall. By the end of 1932, Ford had sold only two Tri-Motors, and his losses in aviation were nearly six million dollars. Prospects for the future were for more losses.[11]

Larger transport planes were slowly replacing the Ford (Fig. 6-7). Eastern Air Transport added the Curtiss Condor (Fig. 6-8) (adapted from a bomber) to their stock, and others like American Airlines also ordered Condors to replace the Fords. The Tri-Motor had served its purpose and was now growing old and obsolete. The government would continue its airmail subsidy to the airlines, and the smaller, slower airlines would soon fall by the wayside.

The American airline network was beginning to jell. It would soon show the world how American free enterprise stimulated unprecedented growth, thanks to Henry Ford. But Ford would not be part of the growth.

6-7 These three Pan American Sikorsky S-40 flying boats are seen over Miami around 1933.

6-8 The Curtiss Condor was an early Eastern Air Transport carrier.

6-9 The 5-AT handled heavy equipment easily. Here eight members of the California Highway Patrol are shown loading their motorcycles into the Ford.

6-10 In 1930, the 40-passenger Sikorsky S-40 flying boats were soon to fly passengers in comfort from Miami to Cuba.

7

The race for the coast

In 1927, the Tri-Motor spelled relative luxury in the sky. There were various cabin arrangements, but the basic airline pattern was nine wicker-type seats, leather cushions, curtains on the windows, and individual reading lamps. Some had dining and sleeper facilities, but these were for the millionaires. American Airlines had experimented with sleeper accommodations on one of their Fords, and apparently people traveling over the vast American Airlines route system welcomed a chance to sleep, even on a Ford.

Crash fever
By 1928, there were 42 airlines flying, and there were 16 fatal crashes on scheduled flights. There had been 24 fatal crashes and 53 planes destroyed since the passage of the Air Commerce Act in 1926. Commercial air travel was becoming popular but not because it was a comfortable or safe way to travel. Business people were beginning to realize Donald Douglas had been right. Where speed was the important element of transportation, the airplane had a definite place in American commerce.

The increasing popularity of commercial aviation was not without a heavy price. As the rickety airplanes climbed into the heavens, the gremlins were hiding in the clouds waiting for them to reaffirm their ownership of the heavens. In 1929, "The Year of the Crashes," it grew worse. The gremlins were winning the battle. The stock market crashed, along with the commercial airlines. That year, record numbers of airliners fell from the sky in fatal crashes, more than in any year before or after. There were 51 crashes, killing 61 people, and the gremlins destroyed 54 more planes. For those people, the adventure of flying had suddenly become fatal, and the industry's poor reputation grew worse. Tony Fokker commented on the crashes: "What astounded me," he said, "was the variety of ways an aircraft would try to kill its pilot and anyone else aboard."[1]

Coast to coast in 24 hours

Commercial air travel had many contradictions, such as the bad press on one hand, and its growing popularity on the other hand. People were finding that it was possible to conduct business between New York and California in a reasonable period, spending only 24 to 30 hours traveling instead of wasting a week on a dusty train.

Time became the all-important element in commercial aviation. If an airline could cut two hours off the coast-to-coast run, it would attract more passengers. One airline, Southern Air Fast Express (SAFE, an acronym not lost on its passengers), had the most complicated arrangement: A train to St. Louis, Missouri; a Tri-Motor to Sweetwater, Texas; a train to El Paso; and another plane to Los Angeles, with intermediate stops for gas and mail.

Soon the race for the coast was on. Before TAT's 24-hour service to California, Universal Aviation Corporation and the Sante Fe Railroad had an arrangement where passengers could go from New York to Los Angeles in 60 hours. Their route was a train to Cleveland, a Fokker trimotor to Kansas City, and a train from there to Los Angeles.

On July 7, 1929, Transcontinental Air Transport inaugurated the first coast-to-coast air-rail service. Now people could go from New York to Los Angeles in close to 24 hours, and with only one ticket. Passengers boarded Pennsylvania Railroad's "Airways Limited" in New York City and arrived in Columbus, Ohio, the next morning. In Columbus, they boarded a Ford Tri-Motor bound for Waynoka, Oklahoma. There were intermediate stops along the way for gas, refreshments, and mail. They included Indianapolis, St. Louis, Kansas City, and Wichita. Arriving at sunset at Waynoka, the passengers boarded the Sante Fe Railroad. The train took them overnight to Clovis, New Mexico. At Clovis they boarded another Tri-Motor. This Tri-Motor stopped at Albuquerque, Winslow, and Kingman, Arizona, before finally reaching Los Angeles. Passengers took the train since there were inadequate navigational aides to fly over the Rocky Mountains at night. This somewhat faster mode of transportation was expensive, too. The one-way fare was $351.94.[2]

The TAT operation was unique. While other airlines carried mail to subsidize their revenue, TAT flew some flights just for people. On some airlines, passengers had been secondary to mail, and they were often "bumped" if the mail was too heavy. TAT provided the amenities that catered to a luxury clientele. Film stars, politicians, and industrialists all favored TAT. Passengers had in-flight meals served on gold-trimmed plates, and each passenger received a gold fountain pen as a souvenir. All the airlines, but TAT in particular, tried to imitate the railroads in providing comfort. Some came close, except none could provide the relaxing clickity-clack of the railroad wheels or eliminate the nerve-wracking, spine-wrenching roar of three engines (Fig. 7-1).

7-1 The Ford at the height of its popularity, flying over New York City.

Air-rail service was just a step in the evolution of commercial air travel. Not only was it costly, but without the mail subsidy, it was a loser. In the first year and a half, the TAT operation lost $2,750,000. The time it saved the passenger was usually not worth the fare, and no one ever rationalized the inconvenience of the plane-train connections. Soon this would pass into oblivion. When cross-country night flying became a reality, all the inconvenient train transfers would look awkward and foolish.

The airlines were slowly realizing that they must change their attitude toward the passengers. A quotation from the 1930 Aircraft Year Book sums it up best: "Only a short time before, he (the air traveler) poked his way into a cumbersome flying suit, buckled himself into a parachute, and waddled out on a dusty field like a penguin, to prepare for each flight. The patience of the uninitiated passenger was tried to the limit. First he stood on one foot, then on the other, waiting for the engines to warm up. Some tried flying in those days. Few found it in keeping with their notion of comfortable transportation" (Fig. 7-2).

The window that opened and closed at 10,000 feet

The act of leaving the ground and defying the laws of gravity upset many people. This, plus the turbulent air playing tricks on the delicate mechanisms of the inner ear, caused many people to be airsick.

7-2 Ford Inflight.

Northwest Airways used Ford Tri-Motors on its routes from Chicago through Minnesota, the Dakotas, Montana, and Washington State. They advertised the Ford Tri-Motor as a plane "with windows that open and close, and complete lavatory facilities." There are some funny stories told about those windows.

Since the Ford Tri-Motor flew below 10,000 feet, it often encountered turbulence. The airlines carried "burp bags" at every seat, but these were usually too small for the needs of most passengers. Windows on the Fords that opened were often welcomed, even at 10,000 feet. Many passengers simply leaned out if they were sick. A certain indignity probably occurred to many passengers, but only one ever recorded his experience. A Qantas pilot, Hudson Fysh, was flying as a passenger on a Tri-Motor in 1930. Fysh was in a back seat in the cabin ("the safest in a crash," he said). He was leaning back, with his window open, and his eyes closed. "A passenger in the front seat was very sick into the usual container," he said. "He threw the container out his window, and I got the full blast in the back seat."

One airline, recognizing the inadequacy of the standard airsick containers, bought several thousand ice cream containers instead. They didn't notice that on the bottom of each container were the words: "Thank you—come again."

Discomfort

The first 4-AT, although a sensation for its day, was admittedly crude compared to today's modern jetliners. The plane had inadequate heaters, for

96 The Fabulous Ford Tri-Motors

example, and passengers would later call it a "flying icebox." It was not particularly attractive, and some called it a "flying washboard." The nickname that has lasted over six decades is the Tin Goose. The noise and vibrations from its engines were overwhelming. To be heard in the cockpit, pilots had to shout. In the passenger compartment, despite what passed as sound proofing and insulation, the noise was, as one passenger described it, "like a hundred gremlins hammering on a barn door with little hammers."

The arrival of cross-country flying did not make flying any more comfortable. Some people described travel on a Tri-Motor as torture. Co-pilots on one airline passed out "comfort packets," which contained chewing gum, cotton, and an ampule of ammonia. The gum was to equalize the pressure in the passenger's ears, the cotton blocked some noise, and the ammonia was to relieve airsickness. Airsickness was so common on the southwestern flights of TAT that someone suggested putting pictures of the Grand Canyon on the bottom of the "burp cups" so people would not miss the scenery.

The primitive nature of the cockpit instrumentation also added to the passenger's discomfort. The early Tri-Motor lacked the instruments that would allow it to fly in any weather, and the cabins lacked the pressurization that would allow the pilots to fly over bad weather. A TAT pilot recalled his early days, when he flew the Fords: "I would get a Ford up to eighteen thousand feet, even twenty-one thousand feet," he said, "and fly reasonably well. Not for long, though. The passengers would pass out." He never explained why he did not pass out.

Weather was especially dangerous to the Tin Goose. The exposed surface controls protruded from the fuselage next to the cockpit. Icing was a problem, too, and often weather forced cancellation of a flight. When this happened, the passengers boarded a train. The TAT pilots contemptuously called TAT, "Take-A-Train" (Fig. 7-3).

When passengers did complete a flight like the Columbus, Ohio, to Waynoka, Oklahoma, run, they'd get off the plane at dusk, weary. Their clothing was often soiled, and their heads throbbed from the constant propeller noise and 117 decibels of sound from the engines. Their bones ached, their nervous systems were like a jumble of skinny wires all sounding different notes, and the exhaust fumes made their noses burn; but they were participating in progress.

Eastern Air Transport Service began passenger service between what is now LaGuardia Airport, New York, and Richmond, Virginia, on August 18, 1930. The daily route covered 310 miles using Ford Tri-Motors. Stops along the route included Camden, New Jersey, Baltimore, Maryland, and Washington, D.C. The service was so popular and grew so rapidly that on December 10, 1930, Eastern added six 18-passenger Curtiss Condors to the fleet. That was the beginning of the end for the Tin Goose.

7-3 Surface control horns on the Ford Tri-Motor were always subject to icing.

Competition for passengers

Passenger service was catching on, and the airlines were trying new ideas to lure the passengers and their money. In January 1929, Universal Airlines equipped its Fokker F10-As with a kitchenette and introduced a meal service on their Chicago-to-Cleveland run. This new service drew more passengers, and while it was innovative for American commercial aviation, Lufthansa had been providing a hot meal service on its Berlin-Vienna Express for several years.

In October of the same year, Transcontinental Air Transport introduced in-flight movies on westbound flights. They were silent (who could hear over the roar of a Tri-Motor anyway?) and included a newsreel and two cartoons. Other airlines followed the example of Universal and offered boxed lunches to lure passengers. It worked; but more passengers meant bigger planes, and more of them.

The industry had been meeting the challenge of more and bigger but not safer planes, and the result was that the equipment inventory in the late 1920s was a jumble of mismatched planes. The scheduled airlines had nine different types of passenger transports. Six of them were single-engine planes that carried six or seven passengers, and three were trimotors: the Fokker F-10A, the Ford Tri-Motor, and the Boeing 80A, which carried from 12-18 passengers. Only the Ford was all metal. In all there were 383 different airplane types certified to 80 different manufacturers in the U.S.

The number of planes on the scene should have been a hint to Henry Ford that something was happening and that he should try something new. He did not. Instead he kept modifying the Tri-Motor and ignoring the competition.

Ford's thoughts went toward two extremes. On one hand, he recognized that for aviation to pay its own way, 100 passenger airplanes were necessary. He also clung to the idea of a "flivver" plane for just one or two people.

His thoughts about 100-passenger airliners stemmed from his belief that commercial aviation relying on passenger income alone would never be economical without mail contracts. A plane with the capacity for at least 100 passengers would offset the loss of mail revenues.

"I am sure we can get a one hundred passenger airplane eventually," Stout said. "From our eleven passenger 4-AT we can build a thirty-passenger plane and then before you know it, we'll build a hundred passenger plane."

Again Henry Ford and Stout were at odds. "The gradual approach is not for me," said Ford. "Instead of us continuing to build small planes from which we would learn nothing, I would rather build a big plane even if it did not fly."[3] Henry Ford got his wish when he built the 14-AT.

There was another multiengine passenger plane operating during this time. The Fokker F-32 was at the time the world's largest land plane in United States airline service. It seated 32 or slept 16 and was clumsy, inefficient, and badly underpowered. Western Air Express and TAT flew them for awhile but withdrew them from service in 1931. Their size and capacity were a hint of things to come (Figs. 7-4, 7-5).

7-4 This Ford was once owned by RCA from 1936 to 1940. A former owner, Gaylord Moxon, installed Boeing P-12 engines. This Ford is now owned by the Hill Country Museum in Morgan Hill, California. Note the B-25 bomber to the left.

7-5 Eventually, TWA got the better design, a Douglas DC-3.

Ford crashes

There were many Tri-Motor crashes (because of overloading, pilot error, engine failure, weather, etc.), but not one ever involved structural failure. Never did a Tin Goose lose a wing or a tail, a tribute to Ford's strict manufacturing standards. Controllable pitch propellers that enabled a pilot to change the angle of the propeller or to stop it (feathering) were not widely used until the mid 1930s. If an engine (or two) failed, the windmilling propeller on the dead engine caused severe turbulence, particularly if it was a wing engine. This turbulence caused several fatal crashes.

In the 1927 Ford Reliability Tour, a Ford Tri-Motor developed heavy vibrations from a faulty propeller. This in turn tore away part of the right side landing gear and damaged the propeller on the opposite wing engine. The pilot cut power to both wing engines but now had a Tri-Motor with just the nose engine operating. He landed the plane without further damage, and mechanics refitted the plane with temporary wooden propellers and repaired the landing gear. The plane went back into the race and finished the 2,584-mile trip. The manufacturer discontinued the propeller that caused the problem. It is not difficult to speculate on what would have happened if the same condition occurred on a wooden Fokker trimotor.

In 1939, a Ford entered the National Air Show at Daytona Beach. During the show, the right engine blew up and fell off the wing. The wing remained intact. Harold Johnson, considered the best of the Ford Tri-Motor pilots, was not the least bit worried. Johnson had done snap rolls at 800 feet and spins at 1,000 feet, and he held the world's record for 27 loops in the six-ton Ford. He landed the plane, and that evening had the nose engine moved to the

7-6 Soon after the F-32 disappeared from the market, the Douglas DC-2 appeared. It would be the real prototype of the future large airliners.

right wing engine location. He then flew the entire exhibition the next day with two engines.[4] Another Ford Tri-Motor lost an engine while crop dusting. The pilot maintained level flight by experimenting with the power settings on the center engine.[5]

The first fatal crash of a Ford Tri-Motor was a Maddux Ford with six persons aboard, enroute to Phoenix from San Diego. An Army Curtiss P-1B tried to loop around the airliner but lost control and crashed into the Ford. The airliner hit the ground upside down, killing all on board. The Army pilot died when his half-opened parachute caught on the tail of his doomed fighter.[6]

The second Maddux Air Lines crash was worse. The plane was on its way to Los Angeles from Mexico when it ran into a squall. The pilot attempted to turn back, but the density of the storm prevented him from realizing he was only a few feet above the ground. His left wing hooked into the ground, and 16 persons aboard died in the crash and fire. The *New York Times* called the crash, "the greatest disaster in American air transportation."[7]

The Tin Goose was strong and durable, but its passengers were not. On August 9, 1931, a Ford Tri-Motor took off from Cincinnati, bound for Atlanta. Minutes after takeoff, the rightside propeller tore loose from the engine. This caused the engine to overspeed and tear loose from the wing. The wing remained intact, and the pilot headed for the only piece of land available, a sandbar on the bank of a river. When the plane's wheels touched the soft, wet sand, the plane ground looped and landed on its back, killing eight people. Ironically, there was no damage to the baggage or the mail.

In contrast, on April 29, 1931, 10 persons survived a serious crash involving a National Air Transport Ford Tri-Motor. The Ford was on its way to Cleveland when the right engine quit. Then the nose engine quit. The pilot attempted to land in a field but overshot the touchdown. The plane tore

The race for the coast

7-7 This Eastern Airlines DC-2 is shown here in 1934.

through some trees, ripped the roof off a barn, had its nose engine torn off, and lost its landing gear. Remarkably, the plane landed upright in the farm yard, scaring the daylights out of the animals, but neither animals nor passengers were injured. The plane was a total loss, and the accident was listed as "cause unknown."

Not all Ford passengers were so lucky. The luxurious Ford of Prince Bibesco of Romania flew into a vulture. The mid-air collision caused the $65,000 Goose to crash to earth, destroying itself and all aboard.

On March 17, 1929, a Ford Tri-Motor owned by Colonial Airways crashed in Newark, New Jersey. There were 15 people on board, and only one, pilot Lou Foote, survived. At the $600,000 negligence trial, Foote

7-8 This Johnson Flying Service DC-2 shared its duties with several Ford Tri-Motors.

102   The Fabulous Ford Tri-Motors

described his experience in the ill-fated Ford. The ship took off properly, but a little sluggishly, he said. He had just completed a right turn at 600 feet when the left motor failed, followed by a drop in the speed of the nose engine. He described trying to glide the aircraft to a safe landing and waking up in the hospital.

From testimony given during the trial, clearly excessive drag had caused Foote to "lose control" of the plane. Experts testified for the pilot. Frank Hawks, holder of the Transatlantic speed record, went on the record saying that Foote had made a proper takeoff and turn. He said, though, Foote may have made a mistake in attempting to make a second turn after the wing engine had failed. Hawks said that Foote should have landed immediately.

The plaintiffs contended that a pilot should not attempt a turn in a Ford Tri-Motor under 1,000 feet. That way, if an engine failed, the aircraft would have enough room to glide safely to a landing. The judge dismissed this argument.

The plane, loaded with Sunday sightseers, had flown the circuit often. An expert witness testified that a full load of passengers (13) was a dangerous

7-9 "Flappers" of the Roaring 20s enjoying tea with the pilot before their flight aboard the Tin Goose.

overload condition. Foote had been carrying 15, including himself. On cross-examination, the witness said that the "extra person had not caused the crash, although the additional weight might have changed the place of landing."

The judge's charge to the jury took more than two hours and stressed that the plane's passengers assumed "all ordinary perils of airplane travel, and all dangers that could not be avoided by the exercise of reasonable care." The jury awarded eight of the plaintiffs a total of $161,000. The headlines read, "the worst accident in the history of heavier-than-air aeronautics in this country." People in the United States were stunned because 14 persons could die so suddenly and so violently in an air crash. (A year before, an equal number of people had died in a Fokker crash, in Rio de Janeiro.) On its front page, the *New York Times* described the condition of the crushed, dead bodies. This had some effect on public opinion, but did not slow the growth of the emerging commercial aviation industry.

8

Ford legal problems

In 1930, the Ford Motor Company ran into a major legal problem with their Tri-Motor. On April 25, 1930, the Junkers Airplane Company of Germany sued the Ford Motor Company in International Court. Hugo Junkers had been the first man to patent an all-metal, low-wing, enclosed-cabin plane in Europe—the F-13, in 1909. (It did not fly, however, until 1919.)[1] A later version of this corrugated-metal, low-wing trimotor was in European service at this time.[2] The suit alleged that the German Ford Motor Company violated the Junker's metal airplane patent by displaying the Ford Tri-Motor in Berlin and other European cities. The suit was a smokescreen intended to obscure more serious issues. The action by Junkers came after a Ford Tri-Motor, sold to the Czechoslovakian government, crashed, killing several important political figures.

The Ford counter-suit in the Prague Trade Court argued that the Fokker and Dornier Airplane Companies used the same patent techniques and were just as guilty. Ford attorneys argued that Junkers' patent should be invalidated. The court refused Ford's request. After a bitter legal battle, the court upheld the Junkers' patent for the construction of airplane fuselages and wings from corrugated metal (Fig. 8-1).

During this period, there were strong anti-American feelings in Europe and in particular, Germany. The Third Reich subsidized the Junkers Airplane Company, and the suit had strong political implications.

Junkers had visited Ford in Michigan, and had also sold several corrugated-metal planes in the United States (Fig. 8-2). Despite similarities between the Ford and the Junkers planes, there is enough evidence to establish that the Ford Tri-Motor was the product of William Stout's original engineering, and not a copy of the Junkers. What motivated Junkers was the

8-1 The Junkers JU-52, Ford's only all-metal competition.

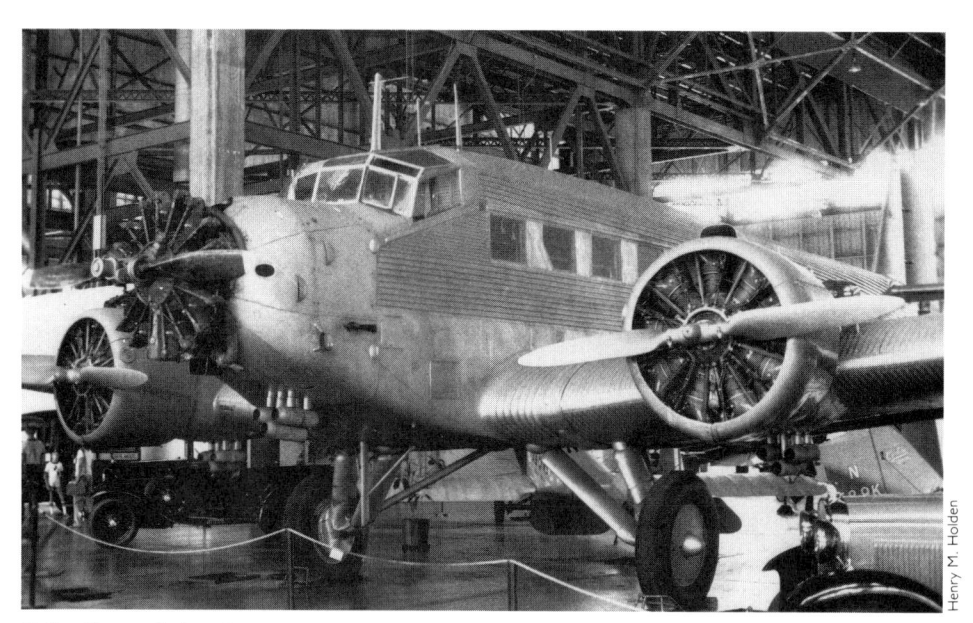

8-2 One of the few remaining examples of the Junkers trimotor, shown here in a museum.

potential mass importation of Fords into Europe. Once Junkers perceived the Ford threat to his market, he acted.

Junkers' perception of a threat was real, too. The Ford Tri-Motor was on a country-to-country European tour. Junkers and the Reich called it American propaganda and once it extended into Germany, Junkers had no choice but to take legal action.

106 The Fabulous Ford Tri-Motors

A Ford Tri-Motor, belonging to the Spanish government's airline, was also seized by the Court to appease Junkers. They later released the Ford, but other European countries did not want to purchase a Ford with the possibility that it might be confiscated. The court's opinion put a stop to Ford's European marketing effort, and the market for Ford in Europe dried up, before it developed.

The flying pioneer

The Ford philosophy was not only to pioneer commercial aviation, but to make inroads into other areas of aeronautics. Many of these inroads would not become widely accepted for years.

All over the world, people and nations were competing for a name in aviation. In 1931, the Japanese Imperial Aviation Society announced a $50,000 prize for any Japanese airman who could fly a Japanese-made airplane nonstop, between Japan and the United States. Although no one ever collected the prize, it did prompt other offers and spurred the competition. Two Americans, Reginald Robbins and Harold Jones of Fort Worth, Texas, decided to fly nonstop from Seattle, Washington, to Tokyo, Japan, in an American-made plane. To do this, they proposed a wild scheme—an in-flight refueling. This had never before been attempted on a transoceanic flight, and naturally many people said it couldn't be done.

The Texans were flying a Lockheed Vega and arranged for the refueling over Alaska. The "air tanker" was a modified Ford Tri-Motor with fuselage tanks to hold 450 gallons of oil and 1,800 gallons of gasoline. On July 31, 1931, the Vega rendezvoused with the Ford over Fairbanks for a refueling. This was to be followed by a second refueling over Nome, Alaska. High winds prevented this refueling, and on August 2, another attempt took place. Plans changed when the crash of two Russian polar fliers caused an all-out air search for the men. The Ford Tri-Motor, part of the search team, ran into bad weather and crashed in the Alaskan tundra. Ironically, the Vega later had its engine replaced with one taken from the wreckage of the Ford. The Wright J5's low fuel consumption better suited the Texans. The Texans did not capture any prize money, but Ford captured another place in the history books. The Ford became the world's first trimotored aerial tanker.

Radio communications

Just as Henry Ford pioneered the commercial airline network, Bell Telephone Laboratories used the Ford Tri-Motor to pioneer the most vital link in that network—communications.

Radios at the time were crude vacuum tube devices, and to be effective, engines required insulating harnesses and grounding around their ignition systems to prevent their radiation from blanketing the radio signals.

Ford engineers worked to improve navigation through fog and storms and began investigating radio as a possible guidance system. Both the Army Air Service and the U.S. Bureau of Standards were also conducting experiments with a radio beacon system when Ford began research on the problem. Ford's efforts paid off on February 10, 1927, when a Ford Tri-Motor, piloted by Harry Brooks, flew from the Ford Airport to Wright Field, in Dayton, Ohio, guided solely by the radio-beacon system. All the major airlines adopted the system, and although Ford held the patent on the radio beacon, no royalties were ever required for the manufacture or use of this system.[3]

The vast Bell Telephone Laboratories (BTL) resources were where most of the major breakthroughs in communications took place, and the Ford Tri-Motor was the vehicle for many of the discoveries. The Ford was BTL's second plane, preceded by a smaller Fairchild. Its pilot, Ray Brooks, a native of Framingham, Massachusetts, and an MIT graduate, specialized in carbon granular research. He was also a World War I Ace. (His Spad, complete with bullet holes, is on display in our National Air and Space Museum.)

Brooks was the first pilot employed by the Bell System, and his job was to organize an air group at Hadley Field, New Jersey. Because BLT had several modifications and systems it wanted in its Tri-Motor, they sent Brooks to Dearborn to supervise the construction of their Ford. Two months later, Brooks had a Ford that by the standards of the day was nothing less than a flying communications laboratory (Figs. 8-3, 8-4). Brooks, at the age of 86, reflected, "The Ford served us well. She was an outstanding aircraft that

8-3 Bell Telephone Laboratories Ford 5-AT-77, NC417H, the flying communications lab shown here over New Jersey in 1930.

108 The Fabulous Ford Tri-Motors

8-4 Interior of Bell Telephone Laboratories Ford. Here it is flying over the S.S. Leviathan. Making antenna tests at the test bench is F.S. Bernhard.

helped us accelerate the science of communications and the development of airborne instrumentation. When we sold it to the Guggenheim Mining Company of South America, they got a pristine-condition, A-1 airplane. I was sad to hear it crashed in some jungle. I helped create that plane."[4]

With Brooks piloting the Ford and doubling as a scientist, the BTL pioneered early radar principles and navigation. Cockpit instrumentation, designed by Bell Labs and manufactured by Western Electric, was years ahead of its time and came in time to help win World War II.

Bell Labs' Ford was the first plane to make radio contact with a ship at sea. On December 22, 1929, it contacted the *S.S. Leviathan*, 200 miles at sea.[5] Earlier, newsmen had talked from the air to the ground, and on to correspondents in London.

The communications equipment developed by the Laboratories performed three functions: it permitted two-way, plane-to-plane communication; it enabled airplane pilots to receive radio-beacon signals and weather information broadcasts; and it allowed communications between four points within the plane itself.

The Laboratories faced two major design restrictions: the size of the aircraft and Federal regulations. The equipment had to be small enough so it didn't add too much weight, and its operation required a minimum of attention.[6]

From these requirements came the 8A aircraft transmitter and the 9A aircraft receiver, a compact, somewhat sophisticated communications package that became standard equipment on the Lockheed, Douglas, and Boeing airliners in the 1930s.

The Ford also helped develop the first accurate radio altimeter and radio direction finders. United, American, and TWA all purchased this equipment when it went on the market.

Brooks logged more than 2,800 hours in the Ford and flew to air shows from Buffalo to Seattle, not to dazzle the crowds, but to display the latest developments in aerial communications. Everyone showed great interest in these displays, since the communications and navigational aids in use were unreliable.

Bell Labs also developed special silencer-type microphones that blocked out the noise of the engines and propellers; something not easy to do in a Ford Tri-Motor. To improve communications, part of the microphone's output was carried back to the receiver so that the pilot could hear his voice over the roar of the engines.

The Labs used the Ford Tri-Motor to experiment with various sources of power for the on-board communications system. Wind-driven generators mounted on the landing gear struts, engine-driven generators, and dynamotors were all tried. Western Electric, the manufacturing arm of the Bell System, made all three available to the airline industry.[7]

Scientists also used the Ford to experiment with airborne antennas. At one point, the Ford carried two trailing wire antennas, an unloaded strut antenna, and a horizontal "V" antenna composed of wires running from the tail to struts on each wing tip. In 1930, Western Air Express adopted the horizontal "V" antenna for its airliners. Other companies opted to use the strut antenna.

In 1930, the Ford became the first "eye-in-the-sky," with a pilot reporting New York City traffic congestion to police headquarters by radiotelephone. There was even talk of using the Ford to chase criminals escaping by car. Scientists used the Ford to develop a special radio receiver that world-speed-record holder Frank Hawks used in his record-breaking, 18-hour, 22-minute Los Angeles-to-New York flight.

In the early days of aviation, successful reception of weather reports and beacon signals was essential for safe flying. If a pilot followed a radio beacon above the clouds to avoid bad weather, the beacon's failure might result in his becoming lost and forced down when he ran out of fuel. If a control tower could radio reports of weather conditions to a pilot, he could try for an alternate field. These were some problems that Bell Laboratories successfully tackled with the Ford. Reliable radiotelephone, radio beacon equipment, and radio altimeters became standard equipment on the commercial airliners that followed the Ford.[8]

The Tin Goose carved a trail across the sky in areas then uncharted. With the help of the Ford Tri-Motor, Bell Laboratories developed aircraft radio equipment that set standards widely adopted by the airlines. The Ford also helped develop instruments necessary for "blind" flying, and the airlines installed them as fast as they became available. In the late 1930s, Bell Laboratories bowed out of aircraft communications research, but the Ford had left its mark. It enabled the United States to enter World War II with an aerial technological edge.

Other uses

The Ford Tri-Motor has many first-time events to its credit. It carried neon signs at night for advertising, searchlights, and even a public address system. Monarch Foods used a Ford Tri-Motor as a portable grocery store to promote its products. Between 1929 and 1931, it flew throughout the midwest displaying the canned foods marketed by Monarch. The walls, above and below the windows, had special shelves so the cans would not roll off while the plane was airborne. With all the seats removed and the canned goods lining every inch of wall space, it might well be called the world's first "flying grocery store."

Another Ford, originally purchased by Maddux, was sold to Transportes Aereos Centro Americanos (TACA) and made history for a mining company. For this operation, the plane underwent a drastic modification. They sealed the cockpit from the cabin and welded a steel ladder beneath the pilot's window. They replaced the cabin windows with corrugated skin, installed a six-hundred-gallon diesel oil tank, and made the cabin water tight. The Ford became the first flying oil can. For several years, it made four flights a day, delivering oil to the La Luz Mine from Alamacamba, Nicaragua.

During her South American migrations for the San Luis Mining Company, another Goose piled up an unsurpassed record of dependability and service. She flew more than 5,376 hours, 4,477 of these without a single incident. During this period, the Ford carried more than 65,000 passengers, 436 tons of mail, and over 47,000 tons of freight. This is an outstanding record for the terrain conditions. The area was mountainous, and the airfield consisted of a single 1,200-foot runway at the bottom of a box canyon, encircled by 4,000-foot walls[9] (Fig. 8-5).

Another Ford gets credit for being the world's first fully equipped aerial

8-5 This is 5-AT-103, NC436H, shown here in 1948 with upgraded Vultee BT-13 engines, controllable pitch props, and cowlings. Note sliding cargo doors added. For a size comparison, that is an old Constellation in the background.

hospital. This Chilean Air Force Ford Tri-Motor, used for medical emergencies, had a complete set of surgical instruments, a sterilizer, scrub sinks, and an operating table. The first flying ambulance was also the first Mobile Army Surgical Hospital.

A Royal Typewriter executive had another unusual use for the Ford. Royal had developed a portable typewriter that they felt was strong enough to withstand a fall from a table. Royal went out and bought a Ford Tri-Motor as a publicity stunt for their product. To prove the strength of this new machine, Royal planned to drop them by parachute from the Ford. The Ford Company installed a special fuselage hatch to replace a window. The interior held over 200 typewriters attached to parachutes.

On August 5, 1927, just two months into its life, the Ford made two aerial drops, one from 250 feet and the other from 700 feet. All the typewriters landed safely, and the advertising campaign was off to a flying start. By September 15, the plane had flown almost 15,000 miles, visited 114 cities in 27 states, and dropped over 13,000 typewriters. A short time later, Royal, deciding they had gotten all the publicity they could from the campaign, ended the parachute delivery system, and sold the plane.

A touch of irony even befell one Ford Tri-Motor. On November 21, 1928, Mrs. Calvin Coolidge, wife of President Coolidge, was at Hoover Field in Washington, D.C., to christen a Pan American Airways' Fokker F-10-A going into service. Unfortunately, the Fokker was not ready for service. The officials, a little frantic, rolled out another plane to act as a proxy. It was a Ford Tri-Motor. The Ford became the first plane to "stand in" for a Fokker, but the Tri-Motor went on to have the last laugh by outliving the Fokker tri-motor.

A silent tribute to the strength and durability of the old Goose sits in the Ford Museum in Dearborn, Michigan. The Floyd Bennett, named for the pilot who flew Admiral Byrd over the North Pole (in a Fokker trimotor), was used by Byrd to fly to the South Pole (Fig. 8-6).

8-6 The Ford Tri-Motor 4-AT-15, NX4542, before takeoff for the Antarctic.

The plane was built as a standard 4-AT, and after it was flight tested, the engineers decided it would not do the job as designed. They modified and increased the wing span, added extra fuel tanks, and installed a Wright Cyclone engine with a three-bladed propeller in the nose. The two Wright 220 hp engines remained on the wings. The engineers installed skis on the plane and flew it to Canada, where they put it through extensive tests in the snow.

All the preparation paid off on November 28-29, 1929, when the plane became the first to carry a man to the South Pole. Even with the additional power from the Cyclone engine, the Ford ran into trouble. The plane could not get over the 10,000 foot "hump" on the way to the pole, and Byrd jettisoned 250 pounds of food to lighten the plane. The plane reached an altitude of 13,000 feet during the trip, which took 17 hours 59 minutes flying time.[10]

Storms forced it down in the Antarctic, and Byrd abandoned the plane in a snow drift. Four years later, he returned to the South Pole, dug the Floyd Bennett out of the snow, filled the gas tanks, and flew it back to civilization.

Texaco No. 1

On December 15, 1927, just six months after the first Ford Tri-Motor rolled out of the factory, the Texas Company accepted its Ford Tri-Motor at the Dearborn Airport (Fig. 8-7). The company intended to use the $48,223 aircraft for another "first." In a press release following delivery, the company said, "With the purchase of this plane, The Texas Company started an

8-7 This is 4-AT-14, NC3443, Texaco No. 1.

aggressive policy to obtain its share of business from the commercial aviation market for petroleum products." Others were now seeing what Henry Ford had seen all along.

The way Texaco planned to market its products was to send the plane on a "tour" of the United States. Between February 21, 1928, and April 1, 1928, *Texaco No. 1* made 24 stops on its tour. It traveled from Michigan through Kentucky, Georgia, Florida, Louisiana, Texas, Kansas, Oklahoma, and Arkansas, attracting larger and larger crowds. The first few stops attracted local newspaper people and a few dozen residents. As word of the Tri-Motor preceded it, the crowds grew. At each stop, the pilot, Frank Hawks gave citizens a ride, and the Ford Tri-Motor's popularity spread. When Hawks reached Little Rock, Arkansas, the crowds had grown to over 10,000. In the words of Hawks, "The reception at Little Rock was without a doubt the most stupendous we have yet encountered."

By May 1928, the company newspaper reported, "The plane is doing at least three things well. It is spreading Texaco good will, creating greater interest in aviation, and giving Texaco airplane products (aviation grade oils and gasoline) severe and practical tests and demonstrating their high quality."

During the tour, Hawks recorded some of his impressions: "The Territory between Jacksonville, (Florida) and New Orleans is anything but inviting to an aviator. I was delighted to be favored with a trimotored plane as my chariot. I never saw so many trees and swampy regions in all my life. I shuddered to imagine all the creepy things that exist in that vast area below."[11]

During a storm, Hawks recorded, "The fury of the storm came upon us in Tennessee. It was raining heavily with a terrific wind coupled with a line of squalls that tossed the large plane around like a leaf. Twice I was thrown out of my seat. It was, without a doubt, the worst storm I have ever encountered."[12] Reports like this were common from pilots flying Fords. The plane

114   The Fabulous Ford Tri-Motors

could take punishment that would have destroyed its wood and fabric-covered counterparts.

A postscript to Hawks' log read, "Everybody is enthusiastic, and especially those who are trying to forward aviation. The Ford Tri-Motor is no doubt stimulating the national aviation program."[13]

On December 22, 1928, *Texaco No. 1*'s tour was cut short. Fog forced Hawks to land on a dirt racetrack in Floresville, Texas. When the weather cleared, he attempted to take off. A crosswind, combined with a muddy field, prevented him from reaching lift-off speed, and the plane crashed through several fences and into a house. The pilot and three passengers walked away uninjured, but the plane was beyond repair.

Just before its unfortunate end, *Texaco No. 1* flew to second place in the 1928 Ford Reliability Tour. This would not have been unusual except that a broken crankshaft in the nose engine forced the pilot to fly the last 200 miles on only two engines.

During the one year and seven days of its life, Texaco's Ford Tri-Motor piled up more than 51,000 miles and carried over 7,200 passengers. More importantly, it helped pioneer the development of aviation-grade fuels and lubricants and advanced the acceptance of the airplane as a modern, safe means of transportation.

Other Fords

The Ford Tri-Motor's versatility prompted its owners to come up with some ingenious uses for the airplane. The Ford was the first large plane to operate on twin floats, thus opening up another market to the plane (Fig. 8-8). With

8-8 This is 5-AT-114, NC9657, at North Beach Airport, New York, in 1934. Note the extra door on the left side. Another rare modification was the floats and wheels combination. This Ford was photographed just before it was sold to Colombia.

twin floats and auxiliary wheels, the Ford could land in the water and taxi up on dry land. It could, for example, leave New York City with hunters and land on an upstate lake (Figs. 8-9, 8-10). The Ford Tri-Motor was also the first plane used for large-scale crop dusting. It could carry 1,400 pounds of arsenate dust and cover acres of land for just a few cents an acre. It was capable of laying a 250-foot-wide path of dust, 20 feet above the trees.

8-9 This TWA Ford 5-AT-69, NC410H, is taxiing into North Beach Airport (LaGuardia) in 1934. Note the nose engine was shut down. This Ford was the first multiengine plane to use twin floats.

8-10 This Ford 5-AT-48, NC8410, originally belonged to Northwest Airways. They sold it to Arctic Airways, shown here in 1936 in Seattle, Washington. Notice the skis on dry land. This plane probably landed on snow and was tied down until the spring thaw.

Diesel-powered Ford

Deeply committed to safe air transportation, Ford engineers continued to search for improvements that would reduce the possibility of serious injury in a crash. In March 1930, engineers installed three 225 hp Packard diesel engines in a Ford Tri-Motor. The diesel engine was safer since there was less chance of a fire in a crash. It did not produce any radio interference and was more economical and reliable. It did have one handicap, however: the diesel engine was heavy and needed more power per bore stroke. The diesel Ford, an 11-AT, did fly, but there was no noticeable improvement. Engineers converted the 11-AT back to a 4-AT and ended development of a diesel powered Tri-Motor.

The military

The Tri-Motor served with distinction and valor around the world. Thirteen flew for the U.S. Army, and nine for the U.S. Navy and Marines. The Dominican Air Force and the Royal Air Force used them. The Royal Australian Air Force flew one, and one survived the Spanish Civil War. The Ford Tri-Motor was part of the famous "Flying Tigers" in China, and like some Americans, served to the end in the air over Bataan. It saw action in Italy and the South Pacific, and it helped the British during the evacuation of Dunkurque.

The outbreak of World War II brought fear to many people on the West Coast. Rumors about invasion spread. When the Army received reports of Japanese ships headed for Nome, Alaska, it reacted quickly. It pressed all the civilian aircraft in the area into service. A Waco, a Bellanca, a Stinson, and two Ford Tri-Motors began ferrying men and equipment into place. Each Ford carried as much as the other planes combined and filled the breech until the DC-3s came to the rescue[14] (Figs. 8-11, 8-12, 8-13, 8-14, 8-15, 8-16).

The Ford Tri-Motor as an investment

In the late 1920s, Ford put a price tag on the 4-AT of $50,000, and when the 5-AT rolled off the line, it sold for $65,000, which during the Depression was a lot of money. An advertisement in 1930 said that the Ford Tri-Motor was built for dependable service (which it did provide) and would "resist depreciation to an unsurpassed degree." That was one of Ford's understatements.

8-11 This U. S. Marine Ford 5-AT-105, RR-5-9205, shown here at VJ-6M Quantico, Virginia, in 1936. The Navy designated their transport planes with the letter "R" in 1931.

8-12 This is 5-AT-95, 31-404, an Army C-4A manufactured in February 1931. Note the wheel pants and cowlings.

8-13 This Navy RR-5-9206, 5-AT-113, is pictured here at the N.A.S. Anacostia. Note the absence of wheel pants on this model.

8-14 This Marine RR-4 manufactured in October 1930 was 5-AT-84, A8840. Note the wheel pants and cowlings.

118 The Fabulous Ford Tri-Motors

8-15 This is a Navy Ford once used by an admiral. Note the stars to the left of the door.

8-16 The Navy still has one Ford Tri-Motor, on display at the N.A.S. Pensacola.

Today Ford Tri-Motor owners are asking over $1 million and getting their price. (*See* page 155.)

The Depression had tragic consequences for many investors, including Ford Tri-Motor owners. Through production improvements, Ford had reduced the 4-AT's price to $42,000 and the 5-AT's to $55,000. As the Depression grew deeper and money dried up, the price of the Ford Tri-Motor fell, too. By 1934, a 4-AT with only 800 airframe hours was selling for $5,000. By 1938, the same Ford was down to $3,950. The September 1937 *Aero Digest* advertised a Ford Tri-Motor for $4,600. In 1939, as the country

Ford legal problems 119

was climbing out of the Depression, the selling price began to rise. *Aero Digest* in March 1939 listed a 5-AT for $13,500.

World War II brought prosperity back to the United States and an increase in the price of the remaining Tri-Motors. The price rose steadily until 1957, when a 5-AT sold for $47,600. The investment in a Ford, like a bottle of fine wine, began to appreciate with age. In 1957, it was just beginning to pay off. The jet age was dawning, and the aviation world was beginning to sense the value of a plane with a large cargo capacity and short take-off and landing ability.

In January 1966, Arizona Airmotive Corporation advertised a Ford for $51,000. In 1968, Island Airlines advertised their 5-AT in *Trade-a-Plane*, for $78,000. At the end of 1968, Jack Adams Aircraft Sales Inc., advertised their 5-AT for $97,400 in *Trade-a-Plane*. In 1976, the asking price of a Ford 5-AT was $1,250,000. A 4-AT went up for sale in 1982 for $900,000. In 1990, the Hill Country Transportation Museum refused a $1 million bid for its Ford Tri-Motor, 5-AT-34. If the trend continues, a Ford Tri-Motor will one day be worth its weight in gold.

End of an era

Ford's Tri-Motor appeared at the right moment in history. It came when America was on the move. It wasn't pretty, and it rattled and shook much like its four-wheeled cousin, but it found its place in the pages of aviation history. The absence of wooden spars, bailing wire, fabric, and glue all helped its acceptance. Lindbergh had made aviation a household word, and that sparked further interest in flying. The Tin Goose was fast in its day. With structural modifications like wheel covers and engine cowlings, it could fly up to 115 mph for 500 miles, and if necessary, land in a cow pasture.

Donald Douglas and his wife were traveling across the country in a Ford Tri-Motor in June 1929 and had the cow pasture experience. The flight had run into a severe thunderstorm, and the pilot made an emergency landing in a small pasture in Youngstown, Ohio.

"The only time I flew on the Ford Tri-Motor where it was the right airplane," said Arthur Raymond, chief engineer for Donald Douglas' DC-3, "was on an inspection trip of the Panama Canal. It had big windows, and flew very slow. It had a good view and was great for sightseeing, but not much else."[15]

Even for people in aviation, flying left something to be desired. When Arthur Raymond and Donald Douglas' general manager, Harry Wetzel, went to New York to visit TWA after the Rockne accident, they traveled by train, though the Fokker trimotor and Ford were available and would have cut three days off the trip. "We took the train for two reasons," said Raymond. "We had a lot of ground to cover, and hundreds of details to lay out and because we really wanted to get there. The ghost of Knute Rockne was still lingering."[16]

During this period, the airlines had seen a sharp increase in the number of accidents, and neither man wanted to become another statistic. To sum up the state of commercial air travel in 1932, it was downright dangerous.

After several days with TWA executives, Raymond said, "Jack Frye asked me to go to Kansas City to work out detailed specifications on our twin-engine proposal." Wetzel went back to Los Angeles by train, but Raymond decided this time it was polite to fly, at least as far as Kansas City.[17]

A Ford Tri-Motor to Kansas City

Raymond had another reason for traveling on a Ford Tri-Motor. The Ford was the major piece of equipment in their inventory. Raymond knew what TWA was looking for: something like the Ford, only better. The trip would radically change Raymond's idea of what to design.

When Raymond boarded the Tri-Motor, he received the usual "comfort pack": chewing gum for the pressure, cotton for his ears, smelling salts if he felt faint, and an airsick cup.

Raymond recalled his flight in the aluminum belly of the Ford. "We bumped along all day at low altitude. It was raining and dark when we landed at Kansas City. As we touched down I got a spray of muddy water on my feet from the S-shaped fresh air ventilators. I knew then why most people took the train."[18]

Franklin Roosevelt flew in a Ford from Albany to Chicago to accept the Democratic Presidential nomination. His dramatic appearance after flying through stormy weather only whetted the public's appetite for faster, safer, and more comfortable aircraft.

While Ford was ruling the highways and skyways, two men, William Boeing and Donald Douglas, were busy designing machines that would put the Tin Goose out to pasture.

The first modern commercial airliner

On February 8, 1933, Boeing introduced the twin-engine 247, the world's first modern commercial airliner (Fig. 8-17). It went into regular scheduled service in June 1933 on United Air Lines' coast-to-coast route. The sleek, smooth-skinned plane was without question the major competitor of the Ford Tri-Motor. Ford had stopped Tri-Motor production, and the new Boeing 247 looked like the heir apparent to the commercial market. The rounded fuselage of the 247 was aerodynamically and acoustically light-years ahead of the Ford. For the passengers, there was heavy insulation in the cabin walls, a hot water heating system, double ventilation, and soft, well-upholstered seats. It was *real* luxury in the skies.[19]

8-17 The Boeing 247, the world's first modern commercial airliner.

For pilots, it offered improvements unheard of just a few years earlier: retractable landing gear, engines that could easily carry the plane over the Rockies, and enough reserve power to carry the plane to a safe altitude if the other failed on take-off. Boeing delivered this plus improved instrumentation and navigational equipment.

Shortly after the introduction of the Boeing 247, Donald Douglas introduced a two-engine plane that offered additional luxury features and a cabin scientifically engineered to reduce noise (Fig. 8-18). The engines, like the Boeing 247, were in the wings with cowlings that improved the aerodynamics of the airfoil. Well-padded seats mounted on rubber shock absorbers also helped reduce the unnerving vibrations (Fig. 8-19).

The Boeing 247 and Douglas DC-2 quickly captured the Ford market, and the addition of the sleeper DC-3s (DST) doomed the Tin Goose. The major airlines soon retired the Fords from the regularly scheduled runs, in favor of the Douglas DC-3. Slowly the Goose faded from the scene, but not completely.

Today the Tin Goose goes on. Over six and a half decades after the first of the Tri-Motors rolled off the assembly line, there are at least 15 left, and five are still airworthy. (See Ford Tri-Motor Survivors—July 1990.) (In 1968, there were 12 flying. Time is slowly taking its toll on the old Goose). Until 1978, two were still flying a regularly scheduled airline route. The rest are either in museums or dismantled awaiting repair (Figs. 8-20, 8-21). The Tri-Motor may be slowly fading, but it has a proud, documented history of reliable, steadfast, and somewhat safe operation. Uncounted miles have gone

8-18 The interior of a DC-3. This improvement came with the DC-1, in 1933. The Ford had taught Donald Douglas what not to put in an airplane.

beneath their wings and hundreds of thousands of passengers have flown in the Tri-Motor.

After retiring from regularly scheduled service, many old Ford Tri-Motors became crop dusters, borate bombers for the Forest Service, or freighters. Many were lost to crashes caused by overloading, or left to rot in a jungle. All the survivors have crashed at least once, and like the legendary Phoenix, risen from their ashes to continue, as if for a special purpose, to educate us, to show us what flying was like so long ago.

8-19 Even the size of the Douglas DC-3 was more impressive than the Ford Tri-Motor.

8-20 This Ford Tri-Motor 4-AT-55 was in the back of a museum waiting for someone with time and money to put it back together again.

8-21 Note similarities in the construction of the Junkers trimotor, shown here in 1986 being restored for Lufthansa's nationwide tour in 1990.

9

The flights of the Phoenix

The strangest story of a Phoenix rising from its ashes came out of Papua, New Guinea. NC-401-H, 5-AT-60, first climbed into the heavens in July 1929 and flew off on an odyssey that would take it around the world from England to North Africa, and to its final resting place in the Owen Stanley Mountain Range, in New Guinea. It would, crash, be rebuilt, crash again, and be rebuilt again and again.

The odyssey began when the Ford Motor Company shipped the Tri-Motor to England, for use as a demonstration model. On December 17, 1930, Ford sold the plane to the Earl of Lovelace. The Earl named his plane the *Tanganyika Star*. On December 28, 1930, the Earl and his party left for Tanganyika, East Africa. The *Tanganyika Star* never reached its destination but crashed near Tripoli. The Earl, injured along with the pilot and other passengers, abandoned the heavily damaged plane for ground transportation, considered by all in the party as a safer means of travel.

The Ford Motor Company shipped the pieces back to England, repaired the plane, and on November 27, 1933, sold it to the British Air Navigation Company (BANCO). They rechristened it *Voyager* and operated the plane in France, carrying passengers between Deauville and Dieppe. BANCO then sold it as a relief aircraft for a small European airline. During these flights, it was a common sight over Sweden.

It then went to Guinea Airways, and arrived in Salumaua, New Guinea, in October 1935. In July 1938, it crashed and sustained heavy damage. Again, they pieced the plane together, and it went back into service in June 1940. The Ford Tri-Motor flew for the next 18 months without a mishap, but then, in December 1941, it crashed again in a landing accident. Once again they repaired the Tin Goose and put it back in service.

The Tanganyika Star gets camouflage

When World War II came to the Pacific, the old *Tanganyika Star* joined the Royal Australian Air Force. On February 6, 1942, it flew back to Queensland, Australia, and flew freight for the next 15 months. In May 1942, it became an aerial ambulance.

In November 1942, the Japanese were inflicting heavy casualties on the Australians in a battle unknown to most people. Those who fought in it knew it as the *Kokoda Trail Campaign*, and the battle would write the history of the *Tanganyika Star* for the next 36 years.

On the morning of November 24, 1942, a civilian pilot for Guinea Airways, Tommy O'Dea, was flying the Goose back from Port Moresby for another load of wounded Australian soldiers. Unknown to O'Dea, the gremlins were about to strike again. The dirt landing strip, carved in the jungle, was also cut in a saucerlike depression surrounded by hills. The mountains in the vicinity were 3,000 to 10,000 feet high. The dirt runway was about 2,000 feet long and under normal conditions would not have been a problem for the Tin Goose. The surrounding hills meant a tricky approach, with limited room to execute a missed approach. It was critical that a pilot make the right approach the first time. O'Dea, a skilled pilot, navigated the mountains and hills and made a normal final approach. The main gear touched the earth, and O'Dea was down and rolling toward the end of the runway. Suddenly a wheel hit a soft patch of dirt and the Goose swerved. The plane headed for a group of people, and O'Dea turned to avoid them. The plane ground-looped and crushed O'Dea in the cockpit. O'Dea was seriously injured, and as a civilian, it would take him more than 20 years to receive compensation for his wartime injuries.

As in any war, men and machines are expendable. The Australian soldiers upended the plane, but it was a wash-out. After removing the salvageable parts and medical supplies, they abandoned the old Goose on the edge of the runway. The war went on without her.

When the war ended, normal life resumed on New Guinea. That is, all life except that of the old Goose. She lay quiet, broken and forgotten, on a deserted landing strip rapidly becoming overgrown by the jungle.

Reflections in a sea of green

Fortunately, the jungle never quite covered the old Goose. Thirty-six years later, an Australian journalist who had flown through the Kokoda Pass many times decided to take a closer look at a metallic reflection he had seen occasionally from the air.

V. T. Sanders flew low over the area and noticed the overgrown but faintly visible tracks of an aircraft in the ground. After almost four decades, the ground-loop marks were still noticeable in the jungle growth. Sanders also noticed a large vine-choked object that resembled an aircraft.

Sanders rented a helicopter and flew back to the old Goose's jungle grave with a camera. After the pilot shut down the helicopter engine, an eerie silence returned to the area. Now that he was closer, Sanders could tell it was an airplane. It was a strange looking plane and he wondered how it got there.

The jungle had grown high around the Tri-Motor. Both wings were missing from the fuselage and lay a short distance away. The port and starboard nine-cylinder Wright radial engines were still in their frames, but the nose engine lay on the ground not far from the fuselage. Sanders wondered if the pilot got out of the crushed cockpit alive.

Inside the cockpit, the wooden control wheels were still in place, although badly weathered. Sanders concluded that the heavily damaged fuselage was probably the result of the crash and someone trying to remove salvageable parts. The wings were in fair shape, but most of the camouflage paint was gone from the fuselage. After 36 years, that is the least one would expect.

Sanders flew out of the mountains and back to his job on the *Post Courier*. He had the photographs of the old Goose developed, and slowly, interest in the long-lost Tri-Motor began to grow. The New Guinea government wrote to the Ford Motor Company and realized from their response that this was a rare airplane in the mountains. After more than 36 years, the Phoenix was about to rise from its ashes again.

The New Guinea government enlisted the aid of the Royal Australian Air Force for this, the last flight of the Phoenix. In 1979, an RAAF Chinook helicopter lifted the old Goose out of her jungle grave, and flew her to the Port Moresby War Museum for restoration (Figs. 9-1, 9-2, 9-3, 9-4).

That is not the end of the story. Soon after arriving at the museum, a question arose about her "real" identity. Bruce Hoy, Curator of the National Museum in Papua, uncovered a disturbing bit of information.

When they dismantled the remains, Hoy's people found the manufacturer's plate read, "Model 5-A-T-B, Serial Number 41, date manufactured January 1929." Hoy had a discrepancy on his hands. Was this Ford 5-AT-60, NC-401-H, or another, 5-AT-41?

9-1 RAAF personnel struggle with the nose engine of 5-AT-60, G-ABHO, in New Guinea, in 1979.

9-2 The RAAF personnel are laying out the parts of the old Goose to see what is still left of her.

9-3 An RAAF Chinook helicopter prepares to lift the Phoenix from its grave of 36 years.

9-4 The Phoenix is flying again, not under her own power but flying nevertheless. Note the drogue chute to stabilize the old Goose.

The flights of the Phoenix

Hoy wrote to William T. Larkins, historian of the Tri-Motor for more than 40 years. Larkins replied to Hoy, "It was good news to hear that the Ford has finally been brought out of Lake Myola. Naturally, the news that it turns out to have a nameplate for 5-AT-41 is even more interesting. The solution to the puzzle is simple; what to do about it is a real problem:

> This story starts with the crash of 5-AT-60, on 21 July 1938. It was so serious that the plane was not licensed for two years. On 28 November 1938, Guinea Airways Ltd. bought 5-AT-41 from Charles Babb, in Glendale, California, who dismantled the plane and shipped the parts to Sydney, Australia. Records show it was never licensed in Australia, and probably used for parts for the wrecked 5-AT-60. Since the nameplate reads 5-AT-41, it is almost certainly the entire fuselage.
>
> In the United States, and in most countries, the legal basis for ownership and identification of an aircraft is by the fuselage, all other parts being components. Thus, the wrecked 5-AT-60 ended its career in July, 1938. Someone should have done the paperwork to show 5-AT-41 imported with a new registration number requested. But since no one did the paperwork, all of the official records show 5-AT-60 continuing to this day.
>
> So now what to do? Forget the whole thing and leave the records as is? Upset the Australian Department of Civil Aviation over a 40-year old historical problem? or quietly explain the real story on a museum plaque?[1]

The museum worked out the dilemma, by clarifying the identity on its plaque.

Island Airlines

Another Phoenix, one that is still flying, is also the oldest Goose in the air, Island Airlines' N7584, 4-AT-38. Until 1977, Island Airlines of Port Clinton, Ohio, used N7584 in a regularly scheduled airline capacity. Island Airlines, billed as the "World's shortest airline," even issued timetables for its 12 stops (Fig. 9-5).

9-5 The oldest Flying Ford in her younger days. 4-AT-38 seen here in 1951.

132   The Fabulous Ford Tri-Motors

As a contrast, the history of Island Airlines is more interesting than the plane itself. Tri-Motor 4-AT-38 went into service in 1929 and went through four small airlines before Milt Herschberger purchased it and founded Island Airways (forerunner of Island Airlines) in 1936. It has been with Island Airways over the years, with a few major airlines leasing it for promotional purposes from time to time.

Herschberger, originally a barnstormer, operated Island Airways with two Fords, 4-AT-38, N7584, and 4-AT-42, N7684. He also added some de Havilland Otters, a Boeing 247, and an assortment of other short-haul aircraft, but the Fords outlasted them all. He operated a flying school, an air taxi, a charter service, and, with the Ford Tri-Motors, a regularly scheduled airline service.

Island Airways flew between four small islands on Lake Erie, Marblehead Peninsula, Put-in-Bay, on Small Bass Island (so named because Commodore Oliver Perry "put in" there after his naval victory over the British fleet in the War of 1812), Middle Bass Island, North Bass Island, Rattlesnake Island, and back to Port Clinton. The distance was about 17 miles, and it took about 45 minutes. Herschberger's two Fords flew the route five times a day, six days a week. The Fords made it possible to travel between the islands even when the frozen lake halted boat traffic.

During its 42-year run as a scheduled airline, the Fords flew the islands' children to the mainland school every day. They also hauled every conceivable kind of freight, from machinery to medicine, lumber, and even corks for the winery, located on Middle Bass Island. It carried close to 75,000 passengers a year and 500,000 pounds of freight and mail.

Not long after World War II, when surplus airplane engines became cheap, Herschberger bought up over 100 N3N engines to keep as spares for the Fords. In 1949, he converted 4-AT-38 and 4-AT-42 to N3N engines.

In the 1960s, Herschberger sold Island Airways to Ralph Dietrick, who renamed it Island Airlines. Dietrick added another Ford to the fleet, 5-AT-11 (NC1629M), and the Fords kept doing their thing.

One of the things the Fords seem to do besides fly is crash. There isn't one Ford flying that hasn't crashed at least once, and Island Airline's Fords were no different.

In August 1972, 4-AT-42, N7684 crashed. Today it is back on the register, owned by Allan T. Chaney. Tri-Motor 5-AT-11, NC1629M, later sold to Scenic Airlines in Las Vegas, also crashed in a landing accident but was rebuilt and later designated N76GC (for the bicentennial). During a wind storm it toppled over, sustaining heavy damage. The San Diego Aero Space Museum is now rebuilding the Ford, hoping to put it on display by 1992.

In 1973, Dave Haberman bought Island Airlines. Haberman, like Island's previous owners, maintained the Fords in their "original factory issue."

In 1977, a pilot took off in 4-AT-38 with only a co-pilot aboard. About a minute into the takeoff, the pilot ran into severe turbulence that interrupted the gravity-fed fuel lines. The result was the failure of all three engines. The Ford Tri-Motors have probably experienced more engine failures than most other airplanes, and like many of its sister ships, 4-AT-38 had multiengine failure at a critical point in the takeoff.

The crash seriously injured the two pilots and destroyed 75 percent of the Ford. When Haberman rebuilt the Ford, he added a piece of equipment unheard of in Henry Ford's day: electric fuel pumps.

Rebuilding 4-AT-38 required a good deal of time, money, and perseverance. Haberman wanted to rebuild it to the exact specifications laid out by the Ford engineers more than a half century earlier. That goal was almost unreachable. Most of the original blueprints were either unreadable microfilm or lost. Through hours of painstaking research in the archives looking at old photographs, and with the help of some dedicated people, 4-AT-38 rolled out of the shop looking exactly as it did when it first rolled off the assembly line. Its reconstruction is accurate, right down to the cowl vents in the nose engine.

While rebuilding 4-AT-38 was costly and frustrating for Haberman, some additional good came from the unfortunate accident. Included in the reconstruction cost was the expense of building the dies and jigs needed to form the parts. These dies and jigs helped rebuild other Fords like 4-AT-58, 4-AT-62, 4-AT-69, and 5-AT-58 at Kal Aero, in Kalamazoo, Michigan. The drawings also helped the people at the Papua National Museum in New Guinea rebuild their Phoenix.

While 4-AT-38 was in the shop undergoing reconstruction, another disaster struck. The FAA revised Part 135 of its regulations. The FAA classifies airplanes according to seat capacity. If a plane seats less than 10, requirements for its operation are different from those aircraft with more than 10 passenger seats.

Haberman's Ford had 15 seats, and the revised Part 135 now required aircraft with more than 10 seats to have weather radar and a cockpit voice recorder. The old Goose would have to carry much of the same equipment as a large commercial jet.

The FAA had literally put Haberman and many other small local airlines out of business. They would have to spend thousands of dollars on sophisticated electronics equipment. Haberman said he would not install the electronics. "The Ford doesn't fly very high, usually at less than 1,500 feet and when weather is approaching, the Ford can be on the ground before it arrives," Haberman claimed.[2]

In March 1980, 4-AT-38 hauled herself back into the air but not on a scheduled run. Haberman discovered that he could still operate the Ford on excursion flights as long as it returned to the same airport.

For a few months, the Ford flew sightseers over the scenic islands of Lake Erie. When the sub-zero weather set in, the tourists disappeared; but Haberman would not put the old Goose in the hangar.

In November 1980, Haberman loaded two extra engines, tires, propellers, and other parts he thought he might need aboard the old Goose and headed south to a warmer climate. Haberman and his Chief Pilot, Harold Hauck, who had flown for Island Airlines since 1951 and had over 14,000 hours in the Ford Tri-Motor, were off to Orlando, Florida. Weather along the way, including an eight-inch snowstorm, turned the two-day flight into four days. The Tin Goose like its fleshy cousin, was still subject to Nature's whims and did not fly very well in a snowstorm (Fig. 9-6).

9-6 The oldest Ford as she looks today.

After six months in Orlando flying excursions over Disney World, Haberman took the old Goose home to Port Clinton (Figs. 9-7, 9-8). She would not fly the dignified, scheduled run she used to, but she'd still give thousands of tourists a taste of flying as the people of the 1920s knew it in a Ford Tri-Motor. Most important, she would not sit in a museum.

Today the oldest Ford is still dividing her time between Port Clinton, Ohio, and Florida, and is now owned by Allan T. Chaney. The Ford Tri-Motor is still not the quiet trimotor jet we know today (Fig. 9-9); she doesn't whisper when she flies, she roars. And why not? 4-AT-38 is the oldest of her breed and she has a lot to say, if we listen.

The flights of the Phoenix

9-7 Every passenger in the oldest Ford was enrolled in the Tin Goose Flying Club and received a Tin Goose membership certificate. Henry M. Holden

9-8 The interior of the oldest flying Ford, 4-AT-38, in a recent picture. Note the wicker chairs have been replaced by steel, and these chairs do have seat belts.

9-9 The American Airlines Ford Tri-Motor is dwarfed by a very distant trimotor relative, the Boeing 727.

10

Other Fords

Early in 1931, the Ford Motor Company announced that customers using their 4-AT Tri-Motors powered by the 220 hp Wright Whirlwind engines could send them back to the plant for "rehabilitation." What came out of the plant was essentially the same plane minus the two-wing engines. In place of the three Whirlwind engines was one 600 hp Wright Cyclone mounted in the nose (Fig. 10-1). Ford engineers said that by doing away with the two outboard engines, the associated fuel tanks, and rigging, the plane's speed would increase 15 to 20 miles an hour and carry 50 percent more payload. An added benefit was an increased cruising range. Ford recommended this modification only for the 4-AT series, since the 5-AT model already carried a larger payload and was faster.

The company denied that it was switching to single-motor aircraft design, something suggested by the trimotor conversion, and stressed its continued support of trimotored aircraft. This later proved true when they announced development of the 14-AT. Yet, Ford did extensive experimental and developmental work on single-engine aircraft, although little came of this.

Surprisingly, few took seriously the invitation to convert their 4-ATs. Only one Ford underwent modification. It became a single-engine freighter and went to Pacific Alaska Airways.

By the early 1930s, Henry Ford had changed his views on the airplane as a deterrent of war. He was now realistic enough to know that nations would use it both as an offensive and a defensive weapon, so Ford converted a Tri-Motor to a bomber. It crashed during tests, and he abandoned the idea (Fig. 10-2). Ford engineers even went back to trying the in-wing mounted engine again. Their tests still showed it to be inefficient and dangerous, and they dropped the idea.

10-1 The single-engine 8-AT-1. This single-engine version of a 4-AT-C went to Pacific Alaska Airways.

10-2 The only Ford Tri-Motor bomber crashed during tests.

Ford now had an assembly line that could fill in six months of normal manufacturing, the entire industry's need for commercial aircraft for the next four years. Yet something had happened. The airlines were not buying Fords in the quantity Ford needed to earn a profit. The structure of the airline industry had changed around Ford, and he had missed the signals. Many early airlines had merged in giant holding companies like United Aircraft and Transport and General Motors. Airline operators were buying their equipment from the parent companies.

During the freewheeling days of the stock market, airlines and manufacturers were merging, some voluntarily and others at the urging of the Postmaster

140 The Fabulous Ford Tri-Motors

General. William Boeing had set up an enormous interlocking directorate. Boeing consolidated smaller airlines and United Aircraft & Transport (UAT), his holding company, had 120 planes in the air covering more than 32,000 miles a day. This made UAT the largest aviation company in the country.[1] By the spring of 1931, UAT was called the "world's largest air transport system."[2] Boeing's subsidiary airlines, BAT, NAT, PAT, and Varney, were large enough to ensure a profit and keep the factory busy with orders. That way, the factory could not sell to an airline outside the organization. This would result in a fatal tactical error.

One forced merger between two reluctant airlines turned out to be very successful. Transcontinental Air Transport and Western Air Express became Transcontinental and Western Air (later Trans World Airlines). American Airways bought up several smaller southwest airlines and a small airline that flew single-engine monoplanes between Atlanta and Dallas/Fort Worth. With Delta Air Service in the fold, American Airways had the southeast link to form a coast-to-coast route. Tony Fokker's American company lost its identity when General Motors formed a conglomerate during the aero-merger boom of 1929. When the merger fever had passed in 1931, no less than 18 airlines ceased being separate operators.[3] The practice became so widespread and abused that the federal government would step in, citing antitrust violations, and break up the holding companies. This, however, would come after Ford had left aviation.

The last of the Tri-Motors

In the early 1930s, Ford began to see the writing on the wall, but he saw two messages. His Tri-Motor sales were dropping, yet the country was turning to mass air transportation. Henry felt there was still a Ford airplane in America's future. Typical of the visionary he was, he began work on a new and larger Tri-Motor. It wasn't the 100-passenger airliner he had talked about, but it was a step in that direction.

The 14-AT, like the 4-AT, was in many ways ahead of its time (Fig. 10-3). It still had three engines, but the third engine was over the center of the wing, not for safety, but to haul the 80-foot monster into the air. It had an incredible 110-foot wing span, and the wing engines were inside the thick wing—the third time Ford engineers had tried this design. The wing presented a smooth, streamlined effect. The engines combined had 2,540 horsepower, more than any other aircraft in existence at the time.

Ford's philosophy of comfort for the traveler was still a major force in the design consideration. The interior of the 14-AT had a smoking compartment immediately behind the pilot's cabin, big, square windows that offered an ideal view, and four main passenger compartments separated by wash rooms, each designed to hold 10 passengers. The compartments, adjoining a

10-3 This giant of the Tri-Motors was also one that never flew.

narrow aisle, were each the width of a double Pullman railroad car. The Pullman-sized seats each converted to a bed. There was a stove and refrigerator in the galley and buttons to call the Flight Escort. The passenger compartments each had thermostatically controlled heat, ventilation, and double insulation to block out the noise.

Before this new trimotor rolled out of the factory, the C.A.A., unsure of its capability, ruled that if the 14-AT somehow managed to fly, it could carry only 10 passengers in its 40 seats. When it did roll out, someone noticed that there was no provision made for steering the plane. The plane, if flown, probably would have crashed. The main landing gear struts were too short, and the wings would probably not get enough angle of attack to generate lift. It was also underpowered. Henry Ford insisted on using the untried Hispano-Suizza engines, instead of the Pratt and Whitney engines originally intended for the plane.

This giant of the Tri-Motors was also the last of them. The Ford plane of the future went to the scrap pile without ever trying its corrugated wings. The 14-AT was a costly failure.

Ford now realized that it was time for him to get out of aviation. The appearance of the 14-AT came at the worst time in the history of aviation, February 1932, when the country was in the depths of the depression. Ford had lost millions on his Tri-Motor designs. There were faster, more streamlined, and more comfortable planes in use. There were others on the horizon, like Boeing's all-metal airliner styled after his B-9 bomber. The Fokker F-32, although inefficient, could carry more people farther in more comfort than a Ford Tri-Motor, and Pan American was using 40-passenger Sikorsky flying boats.

Ford was also carrying a heavy heart. He had lost three young test pilots, one of whom was Harry Brooks, a man he thought of as a second son.

Other trimotors

The Boeing 80A (Fig. 10-4), used exclusively by United Airlines, was a trimotor biplane. It carried 14-18 passengers, cruised at 120 mph, and carried 200 lbs. more in payload than the Ford. It had lacquered fabric covering steel tubing, and like the Ford planes, carried a flight attendant. United, one step ahead of Ford, allowed only registered nurses to fly as "stewardesses." They were the first airline to employ women as flight attendants. This boosted United's popularity because people felt safer with a registered nurse. Boeing built only 12 80As, and its safety record was nearly perfect. Several had ground accidents, but no one ever died in a Boeing 80-A trimotor. In 1934, United bought five used Ford Tri-Motors to replace the Boeing 80As (Fig. 10-5).

10-4 Boeing 80-A.

10-5 United Airlines Tri-Motor.

For a few very good years the Ford Tri-Motor was the Queen of the skies. After her reign, she continued on, surviving crash landings, jungle rot, and every indignity and indifference man could inflict on her (Fig. 10-6).

On September 18, 1932, the *New York Times* reported the Ford Motor Company would stop production of the Tri-Motors. The company said that

10-6 The Stinson Model A flew for Delta from January 1935 to March 1937. The Model A came along after the Fords had been retired by the major airlines, and its life was short because the Douglas DC-3 was on the scene.

the third motor was no longer needed as a safety factor. Over the years, technology had improved to the point where one- and two-engine planes were safe and reliable. Part of their decision, they said, was the high operating cost of the Tri-Motor.

Just weeks before, Ford had asked airline operators to bid on eight completed Tri-Motors. None of the airlines responded. The company eventually sold two to the Navy, and the remaining six were sold at a loss. At this point, many airlines had other manufacturing connections within United Aircraft and Transport, General Motors, and others. The only market left to Ford was the Army or Navy (Fig. 10-6).

Nostalgia

Over the years, to commemorate an anniversary, the surviving Tri-Motors have been dressed up in the colors of the airlines that blazed the commercial trails through the heavens. Each time she flew with the same queenly air, proving that she had never really lost her crown.

In 1949, with World War II becoming a distant memory, the airlines began to look back on their individual histories. TWA was approaching the 20th anniversary of its first transcontinental flight, so they decided to celebrate. They naturally went looking for a Ford Tri-Motor. They found one, dressed it up as the "City of Los Angeles," and sent it on a coast-to-coast tour.

Northwest Airlines celebrated their 30th anniversary with a Tri-Motor leased from Johnson Flying service (5-AT-58). Northwest took their Goose, all spruced up in fresh paint, on a 10-day flight from New York to Seattle, Washington.

In 1962, American Airlines bought 5-AT-39, a plane they originally owned in 1929 (Fig. 10-7). They had it overhauled and then took it on a nationwide tour barnstorming for special publicity events. Over the next few years, the Ford averaged about 400 hours a year (32,000 miles), and about 10,000 people got to ride in a piece of history.

10-7 This is 5-AT-39, taken in September 1961, just before American Airlines purchased it for their 30th anniversary. American Airlines originally owned it back in 1933. This Tri-Motor now flies silently in the National Air and Space Museum in Washington, D.C.

In 1963, TWA could not help but notice the response to American Airlines' publicity. The Ford was attracting crowds everywhere. TWA went out and leased N414H from barnstormer John Louck and jumped into the Tri-Motor business again. They flew a press corps from Los Angeles to Philadelphia to celebrate the 25th anniversary of the Civil Aeronautics Act of 1938. Many newsmen did not like the way the old Ford handled herself in rough weather, and 54 hours later, after they landed, one reporter tried to get even with a blistering article titled, "Last Gasp of the Tin Goose." The reporter called the ride "a desperate journey down memory's airlane."

The old Goose shrugged off the insult like her feathered cousin sheds water and continued to fly more than 15,000 people on Louck's scenic flights.

In 1965, a new adventure was in store for N414H at the New York World's Fair. Youngsters attempted to vandalize her exhibit, but her thick skin showed little more than a few scratches. After all, if the old Goose keeps rising from her ashes, what can a few errant teenagers do to her? Today she sits in Nevada, awaiting a buyer. Her last flight was for a recent German television documentary.

Ford's Tin Goose, like the clouds, continued to roam the skies, even though officially she was "put out to pasture." Old timers called the Tin Goose a, "Faithful old fire horse, unshod and retired to a life of ease and green pastures." A Ford test pilot said 50 years ago, "The end is in sight for the old girl. She was and is a thoroughbred, but her days are almost up." If they could only see her now. The Phoenix keeps rising from her ashes, refusing to die (Figs. 10-8, 10-9, 10-10, 10-11).

Aviation had lost most of its challenge for Henry Ford. He had reached his goal; commercial aviation had arrived. In June 1933, the last Tri-Motor left

10-8 5-AT-58, N8419, was photographed in Seattle, Washington, on its way to Alaska after Northwest sold it.

10-9 5-AT-40, N69905, shown at Orofino, Idaho, in 1952. Originally designated X-ABLF, it flew for CIA Mexicana de Aviacion. After years in Mexico, it returned to the United States only to be lost in Hungry Horse, Montana, in 1953.

10-10 5-AT-80 taken about 1938, at Bettis Airport, Pennsylvania.

10-11 4-AT-58, NC9642, shown here at McCall, Idaho, in 1952. It was almost totally destroyed in a crop dusting accident in 1957.

10-12 This is the Bushmaster 2000, conceived by William Stout in his later years. Construction was started in 1954 but stopped in 1956 when Stout died. Later, construction resumed and the plane was certified.

the assembly line, and the Ford aviation program ended. But not the legend of the Tri-Motor. The accomplishments of Stout, Henry and Edsel Ford, and William B. Mayo inspired the legend, and their leadership helped America grow its commercial wings (Fig. 10-12).

Other Fords   147

The Ford Tri-Motor changed the shape and destiny of world-wide commercial aviation. Like the migrating geese, it found its way from Alaska to South America, and from Europe to Asia, pioneering air travel wherever it went.

With the announcement that Ford would cease airplane production, a significant chapter in aviation history ended; but that was not the end of Ford's involvement in aviation. Ford had dreamed that the airplane would end war. In an ironic twist of fate, Henry Ford and an airplane did help end the bloodiest carnage in history. When World War II broke out, the Ford genius went to work. His factories geared up and began mass-producing Pratt & Whitney engines. There is no question that the United States' ability to mass-produce wartime materials stemmed from Ford's early assembly-line techniques.

When President Roosevelt asked for a 50,000-plane Air Force, Ford said he could mass-produce planes as fast as he did cars. Ford made headlines, planted a winning spirit in the American people, and probably caused the Axis Powers a good deal of concern when he said he could produce one plane an hour. He proceeded to build the largest factory in the world under one roof, and at peak production was turning out a B-24 Liberator bomber every hour.

Ford the movie star

Over the years, Ford Tri-Motors have even starred in the movies. The first was in 1939 in a movie called *Only Angels Have Wings*. The second Tri-Motor starred in a 1950s movie called *The Family Jewels*, costarring Jerry Lewis. The third Tri-Motor flew in the 1984 movie, *Indiana Jones and the Temple of Doom*. This movie has excellent ramp shots and the longest in-flight footage. It also features a simulated crash of the aircraft into the side of a mountain. Another Ford movie appearance was in the 1987 version of *The Untouchables*.

Epilogue

Aviation historians will argue that the most successful commercial airliner in history is the Douglas DC-3, or the three-engine Boeing 727. There were 10,629 DC-3s manufactured and nearly 1,700 Boeing 727s. The success of both airplanes depends on the design. Both were fast in their day, economical, and could carry a good-sized payload. They appeared at the right time in history when a commercial market was in need of a new airplane.

More than six decades ago, another three-engined aircraft appeared in the sky. It too answered a need, and the impact of the Ford Tri-Motor was enormous, considering its legion never reached more than 200. It accomplished what Henry and Edsel Ford set out to prove, that commercial aviation was economically feasible, reliable, and most important, safe. It is hard to believe when you board a sleek Boeing 727 or McDonnell Douglas MC-11 "trimotor jet" that the commercial aviation industry got its start with a slow, noisy aircraft that had only the most basic instruments, primitive creature comforts, and a wrinkled metal fuselage.

When compared to today's modern jet, the Tin Goose is prehistoric. Unlike the prehistoric dinosaur, the idea the Tin Goose stood for did not die and become extinct. It followed a slow, progressive evolution of a species and developed into a more efficient and improved creature. The 3-AT Tri-Motor evolved through the 4-AT, 5-AT, limited versions of 6-AT, 7-AT, and so on through 14-AT. Each was a learning experience, and the lessons were passed on, directly and indirectly, to William Boeing and Donald Douglas. Today, we see the results of that evolution in the sky: the largest, most efficient, and safest commercial airline industry in the world (Figs. E-1, E-2, E-3, E-4.)

E-1 Times and fashions have changed dramatically. Here is a 1928 photograph of a Fokker F-VII of Pan American Airways boarding passengers for Cuba.

E-2 Just 40 years later, there is a dramatic change apparent in the size of the planes flown by Pan American.

150 The Fabulous Ford Tri-Motors

E-3 The Pan Am 747 has a tail that is almost 50 feet off the ground.

E-4 The Fokker F-VII in the lower picture has a tail that is barely four feet off the ground.

Ford aviation milestones

April 23, 1924   The First 2-AT Maiden Detroit, the first all-metal commercial airplane in the United States, rolls out.

April 13, 1925   Ford inaugurates the first regular scheduled airline service (using Stout's 2-ATs for the Ford Motor Company).

July 31, 1925   Ford Motor Company purchases The Stout Metal Airplane Company.

August 8, 1925   Henry Ford announces plans to build a commercial airplane.

August 12, 1925   First Ford airplane reliability tour. Tony Fokker wins in a wood/canvas trimotor.

October 8, 1925   First Ford plane (2-AT) sold to a private individual (John Wanamaker, who also purchased one of the first Ford cars).

January 17, 1926   A mysterious fire razes the Ford airplane factory. Destroyed is the only 3-AT plane along with all the Ford engines and several Wright Whirlwind engines.

February 15, 1926	First Ford airmail route inaugurated from Detroit-Chicago; mileage 252, concurrent with Detroit-Cleveland; mileage 155.
June 11, 1926	Major R.W. Schroeder, Chief Test Pilot for Ford flies the first Tri-Motor Ford monoplane, 4-AT-1, NC1492.
August 7, 1926	Second Annual Ford Airplane Reliability Tour. Two Ford Tri-Motors entered for the first time.
September 3, 1926	Stout Airlines incorporated.
September 18, 1926	Army dirigible RS-1 first airship to moor to the Detroit Airport mast.
February 10, 1927	First long distance airplane flight on which a plane is guided solely by radio (from Ford Airport at Dearborn to Dayton, Ohio and back).
February 15, 1927	Ford airline completes one year of airmail service with 96.6 percent of the scheduled flights over both Chicago and Cleveland routes completed. (Of the remaining 3.4 percent, 2.2 percent is charged to bad weather or lack of mail, and 1.2 percent to flights begun but not completed due to weather conditions. In the first year, Ford mail planes flew 84,000 miles carrying 7,749 pounds of mail.)
March 28, 1927	Airmail mileage increases to 1,436 miles with addition of Detroit to Buffalo route.
May 6, 1927	First air-to-ground telephone call made between William Stout (airborne) and William MacCracken, Assistant Secretary of Commerce for Aviation.
July 27, 1927	The first Ford Tri-Motor goes into service as a scheduled airliner with Maddux Airlines.
November 29, 1929	First airplane, a Ford Tri-Motor, flies over the South Pole (4-AT-15, NX4542).
December 22, 1929	Ford Tri-Motor, 5-AT-7, NC4174, makes first airplane communication with a ship, 200 miles at sea, S.S. Leviathan.
June 4, 1933	Last Tri-Motor to be manufactured, 5-AT-116, NC9659, is flight tested and purchased by Pan Am.

Ford Tri-Motor survivors—July 1990

4-AT-10   N-1077   First Flight 9-10-27
Eugene Frank, Caldwell, Idaho. Dismantled, in poor condition, parts missing. Was N-6077C for many years, now back to its original registration.
Status: under restoration.

4-AT-15   NX-4542   First Flight 3-20-28
Henry Ford Museum, Dearborn Village, Michigan. Byrd's aircraft, first over the South Pole. In good condition.
Status: museum display.

4-AT-38   N-7584   First Flight 9-13-28
Allan T. Chaney, Port Clinton, Ohio. Ex-Island Airlines aircraft rebuilt by Kal Aero after crash 7-1-77.
Status: flying.

4-AT-42   N-7684   First Flight 9-26-28
Listed as Sale Reported to P.O. Box 91, Hebron, OH. (This is another address used by Allan T. Chaney). Back on the register after Island Airlines crash of 8-21-72.
Status: inactive.

4-AT-46   N-7861   First Flight 10-9-28
Naval Aviation Museum, Pensacola, Florida. Painted in Navy Aviation Museum, Pensacola, Florida. Painted in Navy markings as "NAS Pensacola". Built as a model 4-AT-B in Oct 1928; rebuilt by factory as a 4-AT-E in June 1929.
Status: museum display.

4-AT-55   N-9612   First Flight 1-15-29
Dolph Overton, Orlando, Florida. Was listed in 1982 auction of Wings and Wheels Museum aircraft for $500,000. Crashed 6-19-57, added back to civil register in 1980.
Status: under restoration.

4-AT-58   N-9642   First Flight 1-29-29
Kal Aerok Kalamazoo, Michigan. Crashed 6-19-57, added back to civil register in 1980.
Status: inactive.

4-AT-62   N-8400   First Flight 7-6-29
Kal Aero, Kalamazoo, MI. Crashed 7-14-53. One of three ex-Johnson Flying Service wrecks bought by Kal Aero and restored to the Register in 1980.
Status: inactive.

4-AT-69   N-8407   First Flight 8-21-29
EAA Aviation Museum, Oshkosh, Wisconsin. Wrecked in wind storm 6-16-73. Rebuilt and flown 7-20-85. In Eastern Air Transport markings.
Status: flyable, but inactive.

4-AT-65   N-8403   First Flight 5-1-29
Was scheduled to be recovered from the Alaska bush in 1989 by the Alaska Aviation Heritage Museum at Anchorage.
Status: unknown.

5-AT-8   N-9645   First Flight 12-1-28
Evergreen Aviation, McMinville, Oregon. Purchased from Gary Norton d/b/a Henley Aerodrome & Museum of Transportation, Athol, Idaho, 3-90. Sold by the Harrah's auction for 1.5 million. The best Ford in existence in terms of condition and authenticity. Currently active, flying to airshows in Northwest.

5-AT-11   N-9637   First Flight 12-10-28
San Diego Aerospace Museum, San Diego, CA. Was formerly N-76GC of Grand Canyon Air Lines, damaged in a wind storm

at Las Vegas in August 1981. Fuselage currently being completed in basement of museum. Intention is to fly it in about two years.
Status: museum under restoration.

5-AT-34   N-9651   First Flight 3-22-29
Irving B. Perlitch d/b/a Hill Country Transportation Museum and Flying Lady Restaurant, Morgan Hill, CA. Santa Monica 5/90 Auction bid of 1 million refused, plane not sold.
Status: museum, flyable.

5-AT-39   N-9683   First Flight 4-6-29
National Air and Space Museum, Washington, DC. Hanging from ceiling in American Airways markings.
Status: museum display.

5-AT-58   N-8419   First Flight 6-29-29
Kal Aero, Kalamazoo, MI. Being rebuilt after a Johnson Flying Service crash of 8-4-59.
Status: under restoration.

5-AT-60   RAAF   First Flight 7-5-29
(5-AT-41) A45-1
Recovered from Lake Myola by helicopter in Oct 1979. Parts now at National Museum, Papua, New Guinea.
Status: under restoration.

5-AT-74   N-414H   First Flight 9-4-29
Scenic Airlines, Las Vegas, Nevada. Currently inactive, available for sale. Flew from Los Angeles to New York City in Sept 1985 for German television show. Not flying in 1990.
Status: flying, but inactive.

Endnotes

Chapter 1. The fantasy
1. George Vecsey, *Getting Off the Ground* (New York: E.P. Hutton, 1979), 4.
2. All Ford literature and advertisements of the period capitalized "Tri-Motor." All other three-motored aircraft are referred to as "trimotors."
3. This number includes the first 3-AT, and the one 14-AT.
4. William T. Larkins, *The Ford Story* (Wichita: Longo Publishers, 1957), 159.
5. Curtis Prendergast, *The First Aviators* (New York: Time-Life Books, 1980), 128.
6. Peter Collier, David Horowitz, *The Fords An American Epic* (New York: Summit Books), 115.
7. Roger Bilstein, *Flight In America 1900 – 1983* (Baltimore: Johns Hopkins University, 1984), 35.
8. Dixon Stewart, "Aviation Shrine," *All Florida Magazine*, (January 1964), 8.
9. Gay White, St. Petersburg: *St. Petersburg Times*. "Birthplace of Commercial Aviation," 29 November 1953, section G, 2.
10. Bilstein, 15.
11. Douglas Ingells, *Tin Goose*: The Fabulous Ford Tri-Motor, (Blue Ridge Summit, PA: Aero Pub., 1968), 20.
12. John Nevill, "The Ford Motor Company and American Aeronautic Development," *Aviation* Magazine, (15 June 1929): Part III, 27.
13. Bilstein, 15.
14. Ibid., 37.
15. Ibid., 42.
16. David Ansel Weiss, *The Saga of the Tin Goose* (New York: Crown Publishers, 1970), 62.

17. *Smithsonian News Service*, December 1986.
18. *Transactions of the Society of Automotive Engineers*, 1919, "The Airplane as a Commercial Possibility" by Donald W. Douglas April 1919, 444.
19. Ibid.
20. *Scientific American* April, 1920.
21. Carl Solberg, *Conquest of the Skies* (New York: Little Brown, 1979), 103.
22. Alfred Lawson. *A Two Thousand Mile Trip in the First Airliner*. Humanity Benefactor Foundation. no date, 2.
23. *Scientific American*. July 1920.
24. Weiss, 25.
25. Owen Bombard, "The Tin Goose," *Dearborn Historical Quarterly*, (May 1958): 3.
26. William B. Stout, *So Away I Went* (Indianapolis: Bobbs-Merrill Co., Inc., 1951), 29.
27. Weiss, 27.
28. Ibid., 39.
29. Ibid., 42.
30. Grover Loening, *Our Wings Grow Faster* (New York: Doubleday, 1935), 173.
31. Weiss, 45.
32. Ibid., 69.
33. Ibid., 17.
34. Stout, *So Away I Went*, 53.
35. Weiss, 9.
36. John Nevill, "The Ford Motor Company and American Aeronautic Development," *Aviation* Magazine, (15 August 1929): Part V, 32.
37. Ibid., (15 June 1929): Part II, 22.

Chapter 2. The Air Sedan

1. Owen Bombard, "The Tin Goose," *Dearborn Historical Quarterly*, (May 1958): 6.
2. TNT Network documentary, "Reaching for the Stars," 14 February 1989.
3. David Ansel Weiss, *The Saga of the Tin Goose* (New York: Crown Publishers, 1971), 53.
4. William B. Stout, *So Away I Went* (Indianapolis: Bobbs-Merrill Co., Inc., 1951), 323.
5. Weiss, 52.
6. Ibid., 7.
7. Bombard, 7.
8. William T. Larkins, *The Ford Story* (Wichita: Longo Publishers, 1957), 4.

9. Weiss, 69.
10. Ibid., 70.
11. Bombard, 7.
12. Ibid., 8.
13. Ibid.
14. NASM files.
15. Douglas Ingells, *Tin Goose* (Blue Ridge Summit, PA: Aero Pub., 1968), 21.
16. Ibid., 22.
17. Roger Bilstein, *Flight In America 1900 – 1983* (Baltimore: Johns Hopkins University, 1984), 59.
18. NASM files.
19. Kenn Rust, "History of the Airlines," *American Aviation Historical Society Journal*, Winter 1985, Part I, 268.
20. *Aeronautica* Magazine, "Ford Helped Put America On Wheels," June 1963, 4.

Chapter 3. The Kelly Bill

1. John Nevill, "The Ford Motor Company and American Aeronautic Development," *Aviation* Magazine, (15 September 1929): Part VII, 39.
2. NASM files.
3. Roger Bilstein, *Flight In America 1900 – 1983* (Baltimore: Johns Hopkins University, 1984), 86.
4. NASM files.
5. George Vecsey, *Getting Off the Ground* (New York: E.P. Dutton, 1979), 109.
6. Ibid.
7. Owen Bombard, "The Tin Goose," *Dearborn Historical Quarterly*, (May 1958): 9.
8. David Ansel Weiss, *The Saga of the Tin Goose* (New York: Crown Publishers, 1970), 84.
9. William T. Larkins, *The Ford Story* (Wichita: Longo Publishers, 1957), 4.
10. Ibid.
11. Ibid.
12. *Aviation* Magazine, 28 February 1927, photo 399.
13. Nevill, 39.
14. Weiss, 91.
15. Nevill, 38.
16. *Aeronautica* Magazine, "Ford Helped Put America On Wheels," June 1963, 4.
17. Ibid.
18. Larkins, 6.

Chapter 4. The birth of the Tri-Motor

1. Owen Bombard, "The Tin Goose," *Dearborn Historical Quarterly*, (May 1958): 14.
2. "Designing the Ford Tri-Motor," by Tom Towle—Speech at the Annual Meeting, North East American Aviation Historical Society, 28 October 1967.
3. William T. Larkins, *The Ford Story* (Wichita: Longo Publishers, 1957), 7.
4. "Designing the Ford Tri-Motor" by Tom Towle—Speech at the Annual Meeting, Northeast American Aviation Historical Society, 28 October 1967.
5. NASM files.
6. Bombard, 15.
7. "Designing the Ford Tri-Motor" by Tom Towle—Speech at the Annual Meeting, Northeast American Aviation Historical Society, 28 October 1967.
8. *Time* Magazine, 31 March 1967.
9. Larkins, 11.
10. Bombard, 15.

Chapter 5. Launching the Tin Goose

1. David Ansel Weiss, *The Saga of the Tin Goose* (New York: Crown Publishers, 1970), 127.
2. "Designing the Ford Tri-Motor" by Tom Towle—Speech at the Annual Meeting, Northeast American Aviation Historical Society, 28 October 1967.
3. John Nevill, "Ford Motor Company and American Aeronautical Development," *Aviation* Magazine, (15 June 1929): Part I, 18.
4. William T. Larkins, "The Ford Tri-Motor," Profile No. 156, Profile Publications, London, 5.
5. Ford ad in *Aviation Magazine*, 1 July 1929. "Why Ford Monoplanes are made Entirely of Metal."
6. John Nevill, "The Ford Motor Company and American Aeronautic Development," *Aviation Magazine*; (15 August 1929): Part V.
7. Ford advertisement, "The Ford Monoplane is Built for Comfort."
8. Nevill, Part VI, 1 September 1929.
9. Nova TV Documentary, 17 December 1985.
10. Oliver Stewart, *Conquest of the Air* (New York: Random House, 1972), 169.

Chapter 6. Slow growth

1. William T. Larkins, *The Ford Tri-Motor* (Leatherhead, Surry, England: Profile Publications, No. 156, Production list, 1967), 11.

2. Ibid.
3. Kenn Rust, "History of the Airlines," *American Aviation Historical Society Journal*, (Winter 1985): Part II, 274.
4. Ibid, Spring 1986; Part III, 66.
5. Larkins, 11.
6. NASM files.
7. Rust, Part I, 269.
8. Various NASM files.
9. With permission of the author.
10. David Ansel Weiss, *The Saga of the Tin Goose* (New York: Crown Publishers, 1970), 173.
11. Owen Bombard, "The Tin Goose," *Dearborn Historical Quarterly*, (May 1958), 20.

Chapter 7. The race for the coast

1. Richard P. Hallion, *Designers and Test Pilots* (New York: Time Life Books, 1983), 34.
2. Kenn Rust, "History of the Airlines," *American Aviation Historical Society Journal*, (Fall 1986): 174.
3. David Ansel Weiss, *The Saga of the Tin Goose* (New York: Crown Publishers, 1970), 160.
4. William T. Larkins, *The Ford Tri-Motor* (Wichita: Longo Publishers, 1957), 151.
5. Ibid.
6. *New York Times*, 22 April 1929.
7. Ibid., 30 January 1930.

Chapter 8. Ford legal problems

1. *Lufthansa A History*, (Cologne, Federal Republic of Germany: Lufthansa German Airlines, 1975), 13.
2. John T. Greenwood, ed. *Milestones of Aviation* (New York: Hugh Lauter Levin Associates, 1989), 37.
3. Owen Bombard, "The Tin Goose," *Dearborn Historical Quarterly* (May 1958): 20.
4. Telephone interview with author.
5. *Bell Labs Record*, "News Notes," Vol. 8, No. 5, January 1930, 238.
6. Ibid., "Radio Telephone Equipment for Airplanes." October 1930, 7.
7. Ibid., "News Notes." November 1929, 134.
8. Ibid., "Aircraft Radio Receivers." October 1930, 19.
9. William T. Larkins, *The Ford Tri-Motor* (Leatherhead, Surry, England: Profile Publications, No. 156, 1967), 5.
10. William T. Larkins, *The Ford Story* (Wichita: Longo Publishers, 1957), 124.

11. "Log of the Texas Company's New Monoplane," *The Texaco Star* (May 1930), 24.
12. Ibid., 23.
13. Ibid., 32.
14. William T. Larkins *The Ford Story* (Wichita: Longo Publishers, 1957), 125.
15. Arthur Raymond interview with author, 23 March 1987.
16. Arthur Raymond AIAA Address Airporter Inn Irvine, California, 14 March 1983.
17. Ibid.
18. Ibid.
19. This author was aboard the last airworthy Boeing 247 in 1981, and it was by today's standards still luxurious.

Chapter 9. The flights of the Phoenix
1. *Air Enthusiast*, Vol. Twelve, April-July 1980, 50.
2. Fred Haberman, 1981 interview with author.

Chapter 10. Other Fords
1. Harold Mansfield, *Vision* (New York: Duell, Sloan, and Pearce, 1956), 102.
2. R.E.G. Davies, *Airliners of the United States Since 1914* (Washington, D.C.: Smithsonian Institution Press, 1982), 79.
3. Kenn C. Rust, "Early Airlines," *American Aviation Historical Society Journal*, (Fall 1986): Part VII, 162.

Bibliography

Books

Bilstein, Roger E. *Flight In America 1900 – 1983*. Baltimore: Johns Hopkins University, 1984.

Collier, Peter and Horowitz, David. *The Fords: An American Epic*. New York: Summit Books, 1987.

Davies, R.E.G. *A History of the World's Airlines*. New York: Oxford University Press, 1964.

Greenwood, John T. ed. *Milestones of Aviation*. New York: Hugh Lauter Levin Associates, 1989.

Hallion, Richard P. *Designers and Test Pilots*. New York: Time Life Books, 1983.

Ingells, Douglas J. *Tin Goose: The Fabulous Ford Tri-Motor*. Blue Ridge Summit, PA: Aero Publishers, Inc., 1968.

Larkins, William T. *The Ford Story*, Wichita, Kansas: Longo Publishers, 1957.

_____. *The Ford Tri-Motor*. Leatherhead, Surry, England: Profile Publications, No. 156, 1967.

Loening, Grover. *Our Wings Grow Faster*. New York: Doubleday, 1935.

Prendergast, Curtis. *The First Aviators*. New York: Time Life Books, 1980.

Sedgwick, Rhonda Coy. *Sky Trails*. Rapid City, The Story of Cylde Ice. Quarter Circle A Enterprises, 1988.

Solberg, Carl. "*Conquest of the Skies*". New York: Little Brown, 1979.

Stewart, Oliver. *Conquest of the Air*. New York: Random House, 1972.

Stout, William B. *So Away I Went*. Indianapolis: Bobbs-Merrill Co., Inc., 1951.

St. John Turner, P. *Pictorial History of Pan American World Airways*. London: Ian Allen, no date.

Weiss, David Ansel. *The Saga of the Tin Goose*. New York: Crown Publishers, 1970.

Vecsey, George and Dade, George. *Getting Off the Ground*. New York: E.P. Dutton, 1979.

White, Dale and Florek, Larry. *Tall Timber Pilots*. New York: Viking Press, 1953.

Lufthansa A History. Lufthansa German Airlines, Cologne. Federal Republic of Germany, 1975. na.

Magazines (listed by Author)

Bombard, Owen. "The Tin Goose." *Dearborn Historical Quarterly*, May 1958, 20 pps.

Davidson, Budd. "The Last of the Barnstormers." *Air Progress*, June 1986, 71.

Davies, R.E.G. "Pan Am's Planes." *Air Pictorial*, September, October, November, 1967.

Kidd, Stephen. "Ford Tri-Motor—The Golden Goose." *Aviation Quarterly*, 1980, Vol. 6, #2, 2nd Quarter.

Lawson, Alfred. *A Two Thousand Mile Trip in the First Airliner*. Humanity Benefactor Foundation, no date.

McCarthy, Dan. "The Kansas Clipper." *Air Classics*, November 1974.

McGinty, Brian. "Vagabonds with Wings." *American History*, June 1984.

Morgan, Len. "Scenic Airlines." *Flying*", January 1974.

Nevill, John. "The Ford Motor Company and American Aeronautic Development." *Aviation Magazine*, 1 June 1929; Part I, 15 June 1929; Part II, 1 July 1929; Part III, 15 July 1929; Part IV, 1 August 1929; Part V, 15 August 1929; Part VI, 1 September 1929.

Ogden, Bob. "Henry Ford Museum." *Fly Past*, February 1984, No. 31.

New York Times, 30 June 1963, Section 10.

Rust, Kenn. "History of the Airlines." *American Aviation Historical Society Journal*, Fall 1986.

———. "History of the Airlines." *American Aviation Historical Society Journal*, Winter 1985, Part I, II.

_____. "History of the Airlines." *American Aviation Historical Society Journal*, Spring 1986, Part III.

Stewart, Dixon. *All Florida Magazine*, January 1964.

Taylor, H.A. "Ford's Stout-Hearted Trimotor." *Air Enthusiast No. 8.*, October 1978-January 1979.

Towle, Tom. "Designing the Ford Tri-Motor," Speech at the Annual Meeting, Northwest American Aviation Historical Society, 28 October 1967.

_____. "Who Designed the Ford Tri-Motor?" *American Aviation Historical Society Journal*, Fall 1970.

Magazine Articles (not listed by Author)

"Another Ford in Our Future." *Air Classics*, November 1977.

Smithsonian News Service, December 1986.

Transactions of the Society of Automotive Engineers, 1919.

Scientific American, April 1920.

Scientific American, July 1920.

"Flying High in Vintage Fords." *Business Week*, 7 January 1967.

"Log of the Texas Company's New Monoplane" *The Texaco Star*, May 1930.

Time Magazine, 31 March 1967.

"Ford Helped Put America On Wheels." *Aeronautica* Magazine, June 1963, 3.

"Bill Stout." *Fortune*, January 1941.

"Bushmaster 2000 International Tour Set." *Aviation Week and Space Technology*, 26 September 1966.

"Colossus of the Caribbean." *Fortune*, April 1931.

"Detroit da Vinci." *Saturday Evening Post*, 7 December 1940.

The New York Times. Various Henry & Edsel Ford interviews, 1924–1971.

"The Bushmaster." *Aviation Week and Space Technology*, 18 November 1968.

Bell Labs Record. "Radio Telephone Equipment for Airplanes," October 1930.

_____. *Aircraft Radio Receivers*," October 1930.

_____. "General News Notes," January 1931.

_____. "General News Notes," November 1929.

_____. "The Laboratories Take to the Air," May 1928.

"The Bushmaster." *Air Progress*, August 1979.

St. Petersburg Times, 29 November 1953.

Aviation Magazine, 28 February 1927.

Air Enthusiast, Vol. Twelve, April-July 1980.

Other

"The First Fifty Years of Pan Am." The story of Pan American Airways, Inc. from 1927 – 1977, Pan Am publication 1977.

Arthur Raymond AIAA Address, Airporter Inn Irvine, California, 14 March 1983.

Sanders, V.T. "The Gaunt Ghost." *The Post Courier*, 1 October 1979.

Pacific Islands Monthly, "Life and Death of A Trimotor," December 1979, 39.

Index

14-AT Tri-Motor, 141-142

2-AT Air Transport, 26, 27-29, 43, 44, 46, 47, 52-53, 55, 60, 63, 80, 81
   first private buyer, 48

3-AT Air Transport, 52-57, 60, 64, 65

4-AT Tri-Motor, 63-80, 83, 99, 113-114, 124, 133, 134, 135, 147, 149
   advertising the Tin Goose, 69-70
   Alclad aluminum fuselage, 78
   assembly line production, 78
   braking system, 75
   cockpit design, 67, 74-75, 76-77
   comfort features, 72-73
   construction techniques, 70-71
   controls, indicators, 74, 75-76
   cruising speed, 67
   engines, Wright engines, 67, 139
   exterior design, 77
   fuel capacity, 77
   handling and performance characteristics, 66-67, 72
   interior, 33, 74, 136
   life-expectancy of aircraft, 69-70
   maintenance requirements, 73-78
   navigation aids, 72
   night flight, 73-74
   pilots, 67
   rollout and maiden flight, 67-68
   rust-proofing, 70-71
   safety innovations, 69-70
   seaplane adaptations, 70-71
   simplicity of design, 73-78
   size, 64-65
   stressed-skin construction, 78-80
   two-pilot configuration, 75
   visibility, 76
   weight vs. stall speed improvements, 65-66
   windshield modification, night flight, 73-74
   wing designs, 64-65, 80
   Wright engine, 78

5-AT Tri-Motor, 87-89, 91, 99, 1112, 115-118, 129-132, 133, 134, 145-147, 149
   engines, Wright engines, 139

8-AT-1, 140

A

accidents, 40-42, 45, 69, 82, 93-94, 100, 120-121, 133
advertising the 4-AT Tri-Motor, 69-70, 111
Aerial Age magazine, 15
Aero Digest magazine, 119, 120
aerobatics, 5
Aeromarine Airways, 38
Aeromarine Airways flying boat, 38
Aeromarine flying boats, 14
Aeromarine Plane & Motor Co., 56
Air Commerce Act, 43, 86, 93
Air Mail, Britain, 14
Air Mail, U.S., 14, 31, 32, 34, 37-49, 81, 85
   first commercial airmail contract, 48-49
   Ford Air Transport Service, 48-49
   Ford Motor Company airline, 44-45
   hazards of flight, 40-42
   inaugural route, 1918, 38
   Kelly Bill of 1925, 37
   navigation aids, 41
   night flying, 40
   reliability tours, 45-48
   trimotor-type aircraft proposed, 40
Air Pullman, 26
Air Sedan, 23-36
   2-AT Air Transport rollout, 27-29
   Ford Air Transport Service begins, 33-34

Air Sedan (*cont.*)
  Ford takeover of Stout Company, 27
  fund-raising campaign, 23-25
  Stout Metal Airplane Company formation, 25-27
Air Truck, 26
air-rail service, 94-95
Aircraft Board, World War I, 16, 17
Aircraft Development Corporation, 31
airline development, 6, 9, 14, 15, 20, 21, 28-29, 34-36, 38, 40, 43-45, 48, 69-72, 93, 140-141, 144-148
  air-rail service, 94-95
  coast-to-coast travel, 94-95
  comfort features in airplanes, 72-73
  comfort features, 95-96
  competition for passengers, 98-99
  discomforts of early transcontinental flight, 96-97
  modern airliner development, 121-123
  passenger capacity, 99
  reliability tours, 45-48
airliner development, 121-123, 141-142
airport development, 31
airsickness, 95-96
airworthiness certification, 43
Alclad aluminum fuselage, 4-AT TriMotor, 78
all-metal construction, 18-19, 78-80, 105
Aluminum Company of America, 18-19
aluminum fuselage, 4-AT Tri-Motor, 78
American Airways, 5, 82, 90, 93, 110, 141, 145
American Railway Express Company, 12
Anzani engine, 57
Arizona Airmotive Corp., 120
Army 5-AT-D, 84
assembly-line production, 10-11, 61, 78, 90
Auken-Ford Flying Machine, 5-6

automotive industry development, 3
aviation industry development, 5-6, 11, 12, 24-25
Aviation magazine, 18
aviation milestones, Ford Company, 153-154

B

Back Door Gang, 55
barnstorming, 20
Bat Wing design, 17-18
Bell Telephone Laboratories, 107-111
Bellanca, 117
Benoist Type XIV, 6, 9
Benoist, Thomas, 9
Bibesco, Prince, 84, 102
Bleriot, Louis, 3
Boeing 247, 74, 121-122
Boeing 727, 137, 149, 151
Boeing 80A trimotor, 99, 143
Boeing Air Transport, 12, 85
Boeing Airlines, 141
Boeing B&W, 12
Boeing B-1, 39
Boeing Monomail, 80
Boeing, William, 3, 11, 24, 38, 121, 141, 149
bomber, Tri-Motor design, 140
Book of Directions, 41
braking system, 4-AT Tri-Motor, 75
Braniff Airways, 82
Briggs, W.O., 43
British Air Navigation Company (BANCO), 127
Bromma-Stockholm airport, 31
Brooks, Harry, 57, 108, 142
Brooks, Ray, 108, 110
Bureau of Air Commerce, 84
Bushmaster 2000, 152
Bushnell, David, 15
Byrd North Pole Expedition, 46, 63, 64, 113
Byrd South Pole Expedition, 113

C

C-2 Larson Liner, 14
Chamberlain, Clarence, 33
Champion Spark Plug Company, 18

Champion, Albert, 24
Chaney, Allan T., 133, 135
Chang Hsueh-Ling, 84
Chicago Tribune newspaper, 15
Chinook helicopter, salvage operations, 129, 131
Chrysler, Walter B., 24
City of Los Angeles, 144
coast-to-coast travel, 94-95
cockpit design, 74-75, 76-77, 97
Cody, Mable, 20
Colonial Air Transport, 12
Colonial Airways, 102
comfort features, 72-73, 95-96
controllable-pitch propellers, 100
controls, indicators, 74, 75-76
Coolidge, Mrs. Calvin, 112
Cugnot, Nicholas, automotive inventor, 3
Curtiss Aeroplane & Motor Company, 49
Curtiss Condor, 90, 92, 97
Curtiss JN-4 Jenny, 11, 16
Curtiss OX-5 engines, 25-26
Curtiss P-1B, 101
Curtiss Pusher Loop Model, 8
Curtiss, Glenn, 2, 5, 46

D

da Vinci, Leonardo, early designs, 1
Dayton-Wright Co., 56
de Bergerac, Cyrano, 1
de Havilland DH-4, 11, 40
de Havilland GH-4 observation plane, 26
de Havilland Otters, 133
Delta Air Service, 141
Delta Airlines, 87
Depression era, 90-91, 119, 120
Detroit Aviation Society, 45
Detroit Free Press newspaper, 27
Detroit Street Railway Company, 85
Detroit Times newspaper, 42
diesel-powered engines, 116-117
dirigible patents, Ford, 20
Dodge, Horace, 24
Dominican Air Force, 117
Dornier Airplane Company, 105
Douglas Cloudster, 35

170   Index

Douglas DC-1, 74
Douglas DC-2, 101, 102, 122
Douglas DC-3, 12, 13, 14, 48, 83, 100, 120, 122, 124, 149
 interior, 123
Douglas Mailplane, 39
Douglas World Cruisers, 30, 32
Douglas, Donald, 3, 12-14, 18, 35, 38, 120, 121, 122, 149
duralumnin fuselage, 4-AT TriMotor, 78

E

early attempts at flight, 1-6
Eastern Air Transport, 82, 90, 97
 Curtiss Condor, 92
 Ford Tri-Motor, 4
Eastern Airlines DC-2, 102
Edison, Thomas, 48
Egyptian artifacts depicting flight, 1
executive Ford Tri-Motors, 84
exterior design
 4-AT Tri-Motor, 77
 5-AT Tri-Motor, 89

F

Family Jewels, The movie, 148
Fansler, paul, 6
fatalities, 9, 36, 40-42, 45, 93-94, 100-104, 120-121
ferrying missions, 117
Field, Marshall, 24
fire at Ford Aviation plant, 1926, 55
Firestone Co., 84
Firestone, Harvey, 24
Fisher brothers, 24
Fisher, Fred, 43
flight attendants, 34, 142
Flight Escorts, 34, 142
Flivver development, 57-58
floatplane conversions, 115-116
 5-AT, 115, 116
Florida Airways, 43, 48
 2-AT Air Transport, 28
Flyer, Wright brothers airplane, 2
Flying Tigers, 117
Fokker Airplane Company, 105
Fokker DR-1 tri-wing, 45
Fokker F-10A, 99, 112
Fokker F-32, 99
Fokker F-VII trimotor conversion, 45, 46, 63, 70, 71, 150-151
Fokker F10-A airliners, 98
Fokker, Tony, 3, 45, 60-61, 81, 94, 141
Foote, Lou, 102-104
Ford Air Transport Service, 33-34, 48-49
Ford Airport, 30-31, 33, 46
Ford Motor Company, 3
Ford News newsletter, 31
Ford Reliability Tours, 88, 100, 115
Ford Story, 89
Ford, Edsel, 5-6, 11, 24, 26, 27, 30, 34, 42, 43, 45, 46, 53-56, 61, 147, 149
Ford, Henry, 3, 26, 30-34, 42, 43, 44, 46, 51, 53-57, 61, 64, 67, 69, 71, 81, 85, 87, 99, 107, 139, 141, 147, 149
freight hauling, 5, 12, 34, 81, 82, 111, 112
Frye, Jack, 121
fuel capacity, 4-AT Tri-Motor, 77
Fysh, Hudson, 96

G

General Motors, 140, 141, 144
Glenn L. Martin Co., 56
Glidden Tours, automobiles, 48
Glover, W.I., 48
Guggenheim Mining Company, 109
Guinea Airways, 128
Gulf Coast Airlines, 38

H

Haberman, Dave, 133, 134, 135
Handley-Page bombers, 12
Hawks, Frank, 103, 114-115
heating systems, cabin heat, 96-97
Henderson, Paul, 40
Herschberger, Milt, 133
Hicks, Harold, 54, 57, 58, 59, 60
Hispano Suizza engines, 17, 18, 25-26, 142
Holley, George, 43
Hoppin, Glenn, 19
horizontal stabilizer, 4-AT TriMotor, 77
hospital planes, 112
Hoy, Bruce, 129, 132
Hunsaker, Jerome, 40, 41

I

in-flight refueling, 107
Indiana Jones and the Temple of Doom movie, 148
instrumentation, 110-111
interior
 4-AT Tri-Motor,73, 74, 136
 5-AT Tri-Motor, 88
 Douglas DC-3, 123
investment value, 117-120
Island Airlines, 120, 132-133

J

Janus, Roger, 9
Janus, Tony, 9
Japan-U.S. trans-Pacific flight, 107
Japanese Imperial Aviation Society, 107
John Wanamaker Co., 48
Johnson Flying Service, 144
 DC-2, 102
 Ford Tri-Motor, 66
Johnson, Harold, 100-101
Jones, Harold, 107
Josephine Ford, 46
Junkers Airplane Company, 63, 105
Junkers JU-52 , 106, 125
 instrument panel, 76
Junkers, Hugo, 3, 17, 25, 78, 105, 106

K

Kelly Bill of 1925 (*see also* Air Mail, U.S.), 37
Kettering, C.F., 24, 43
Knauss, Stanley, 19, 53
Kokoda Trail Campaign, 128
Koppen, Otto, 57-59, 60

L

L-4 Larson Liner, 14
Larkins, William T., 89, 132
Larson, Alfred W., 14
Law, Ruth, 8, 10
Lee, John, 58, 59, 60
legal problems, 105
Lenoir, Etienne, automotive inventor, 3
Lewis, Jerry, 148
Liberator bombers, 5
Liberty engines, 11, 16, 26, 34, 41, 52-54

Lindbergh, Charles, 86-87
Lockheed Vega, 107
Loening, Grover, 17
Louck, John, 145
Lufthansa airlines, 79

M

Maddux Air Lines, 81-82, 101
Maddux, John, 82
Maiden Dearborn I, 33, 36, 43, 48
Maiden Dearborn II, 34, 47
Maiden Detroit 2-AT, 28, 31, 33
maintenance requirements, 4-AT Tri-Motor, 73-78
Manning, Leroy, 48
Martin MB-1 bomber, 18
Martin, Glenn, 2, 38
Mayo, William B., 19, 20, 24, 30, 32-33, 40, 42, 43, 51, 56, 57, 59, 60, 67, 68, 81, 89, 147
McDonnell Douglas MD-111, 149
McDonnell Aircraft Company, 56
McDonnell, James, 56, 58, 59
mid-air collisions, 82
milestones, Ford aviation, 153-154
military service, 117, 128
Monarch Foods, 111
movie roles for Tri-Motors, 148
movies on board airliners, 98

N

National Advisory Committee for Aeronautics (NACA), 27
National Aeronautics and Space Administration (NASA), 27
National Air Races, 47
National Air Reliability Tour, 45-48
National Air Show, 1939, 100
National Air Transport (NAT), 9, 12, 82, 101
navigation aids, 41, 72, 94, 109
NC-4, 11
New York Evening Post newspaper, 15
New York Telegraph newspaper, 14
New York Times newspaper, 35, 73, 87, 101, 104, 143
New York Tribune newspaper, 34
New, Harry S., 44
night flying, 40, 111
  4-AT Tri-Motor, 73-74

Northrop Gamma, 79
Northrop, Jack, 80
Northwest Airlines, 82, 96, 144
  Ford Tri-Motor, 83

O

O'Dea, Tommy, 128
Only Angels Have Wings movie, 148

P

Pacific Alaska Airways, 139
Pacific Marine Airways, 38
Packard diesel engines, 116-117
Packard Motor Car Company, Aircraft Division, 16
Pan American Airways, 81, 82, 112, 142
  Boeing 747, 151
  DC-3, 13
  Fokker F-VII trimotor, 71, 150
  Ford Tri-Motor, 4
  Sikorsky S-40 flying boats, 92
passenger capacity, 99
patent infringements, 105
Pennsylvania Railroad "Airways Limited" service, 94
Phiel, A.C., 6
Phoenix Fords, 127-137
Pitcairn Airways Inc., 43
Pitcairn Mailwing, 33
Pittsburgh Airways, 82
Port Moresby War Museum, 129
Post Courier newspaper, 129
Pratt & Whitney engines, 142, 147
pressurization in cabins, 97
production line, 82
propellers, controllable-pitch, 100
Prudden, George, 19, 27, 51, 58, 60
Pulitzer Airplane Speed Classic, 45

Q

Quimby, Harriet, 7

R

radial engines, 41, 51, 53
radio communications, 107-111
Rand, Marcell N., 84
Raymond, Arthur, 120-121
reliability tours, 45-48, 88, 100, 115

restoration efforts, 127-137
Robbins, Reginald, 107
Rockne, Knute, air fatality, 120-121
Rohrbach, Adolph, 78-80
Romanian Federation Aeronautica Internationale, 84
Roosevelt, Franklin D., 121, 148
Royal Air Force, 117
Royal Australian Air Force, 117, 128, 129
Royal Typewriter, owners of TriMotors, 84, 112
Russell, Harry, 47
rust-proofing, 4-AT Tri-Motor, 70-71
Ryan Airlines, 38

S

safety, 40-42, 45, 48, 69, 93-94, 100-104, 120-121
sales figures, 89-90, 119-120
salvage and restoration efforts, 127-137
San Luis Mining Company, 111
Sanders, V.T., 129
Sante Fe Railroad, 94
Scenic Airlines, 133
Schroeder, Shorty, 52, 54, 60, 67, 68
Scientific American magazine, 14-15
seaplane adaptations, 4-AT TriMotor, 70-71
selling prices, 119
Shenandoah airship crash, 51
Sikorsky S-40 flying boat, 91-92, 142
Smith, Jimmy, 55
Southern Air Fast Express (SAFE), 94
Spirit of St. Louis, 86-87
St. Petersburg-Tampa Airboat Line, 6
stall speed improvements, 4-AT Tri-Motor, 65-66
Standard Oil Company, 82, 84
Stinson airplane, 117
Stinson Model A trimotor, 144
Stinson Model T trimotor, 63, 64
Stinson, Eddie, 21, 25, 31
Stock Market crash of 1929, 90-91
Stout Air Service, 84-86

Stout AS-1, 25
Stout Engineering Laboratory, 19, 21
Stout Metal Airplane Company, 25-27, 31, 58
 Ford takeover, 42-43
Stout, William B., 15-21, 23-36, 40, 42, 43, 51-60, 67, 75, 84-86, 99
Stout-Ford merger, 42-43
Stranahan, Robert, 18, 23
stressed-skin construction, 4-AT Tri-Motor, 78-80
surviving Ford Tri-Motors (see also Phoenix Fords), 122-123, 132, 155-157
Swallow, 46

T

Tanganyika Star (see also Phoenix Fords), 127-128
Templehof Field, Berlin, 33
Terry, James, 73
Texaco Co., 84
Texaco No. 1, 113-115
Tiffany trophy, 45, 47
Timken, H.H., 84
Tin Goose Flying Club certificate, 136
Tomlinson, D.W., 72
torpedo plane, 19, 21
Towle, Thomas, 52, 53, 55, 56, 58, 59, 60, 64, 65, 67, 68, 81
Trade-a-Plane magazine, 120
traffic control airplanes, 110
Trans World Airlines, 141

Transcontinental Air Transport (TAT), 82, 87, 94-95, 98, 141
 Ford Tri-Motor, 4
Transportes Aereos Centro Americanos (TACA), 81, 111
Travelair, 29, 46
Triad land-sea plane, 8
trimotor designs, 40, 41, 46, 51-61
 3-AT Air Transport, 52-57
 Boeing 80A, 143
 fatigue cracks, 52
 Flivver development, 57-58
 Fokker F-VII conversion, 45, 46, 63, 70, 71
 low-lift wings, 52
 radial engine development, 51
 steel engine mounts, 52
 Stinson trimotor, 63, 144
 test flights, 60
 Towle Tri-Motor, 58-61
 wing design, 60
TWA, 110, 121, 144, 145
 Douglas DC-3, 100

U

U.S. Army 5-AT Tri-Motor, 118
U.S. Marine 5-AT Tri-Motor, 118
U.S. Marine Ford 5-AT, 117
U.S. Navy 5-AT Tri-Motor, 118-119
United Air Lines, 82, 110, 121, 143
United Aircraft and Transport, 85, 140, 141, 144
Universal Airlines, 98
Universal Aviation Corporation, 94
Untouchables, The movie, 148

V

Van Auken, Charles, 5-6
Verne, Jules, 1
visibility, 4-AT Tri-Motor, 76
Voyager (see also Phoenix Fords), 127-128

W

Waco, 46, 117
Waldon, Sidney, 43
Wanamaker, John, 48
Warner, Edward P., 31, 32
Wasp engines, 43, 87-88
weather factors, 97, 110-111
Western Air Express, 12, 99, 110, 141
Wetzel, Harry, 120-121
wing designs, 60, 80
 3-AT Air Transport, 53-54
 4-AT Tri-Motor, 64-65
 5-AT Tri-Motor, 88-89
 Bat Wing, 17-18
World War I aircraft production, 6, 16-18
World War II aircraft production, 5, 10-12, 147-148
World War II military use, 117, 128
World's Fair exhibit, 1965, 145
Wright brothers, 2
Wright Engine Company, 51
Wright engines, 53, 67, 68, 78, 113, 139
Wright J5, 107
Wright Model B, 10

Other Bestsellers of Related Interest

THE BOEING 247: The First Modern Commercial Airplane—Henry M. Holden

Now, you can learn all about the revolutionary Boeing 247—the first modern air transport to prove that commercial air travel could be safe and efficient. You'll trace the development of the Boeing 247, and you'll relive William Boeing's struggle to implement unique designs, the public's reception to the aircraft, its commercial and military uses, and its impact on aviation. Finally, you'll take part in the races and record-breaking performances that marked the Boeing 247's short but impressive career. 184 pages, 109 illustrations. Book No. 3593, $14.95 paperback only

THE DOUGLAS DC-3—Henry M. Holden

This colorful volume traces DC-3 development and production, its varied commercial uses, and its innovative safety features. As you follow this fascinating history, you'll discover the dramatic, human side of the DC-3 legend, trace the evolution of the airplane and discover the vital contributions it made to the post-World War II aviation boom. You'll also find never-before-published facts and photographs. 224 pages, 121 illustrations, 8 full-color pages. Book No. 3450, $16.95 paperback only

THE LUSCOMBES—Stanley G. Thomas

Beginning with the conception of Don Luscombe's famous Monocoupe in the 1920s, this book describes the creation of the Luscombe Airplane Corporation, the development of such important Silvaire ancestors as the Phantom 1 and the Model 4, and the design and production of the Model 8 through final rollout in 1960. You'll learn about the Silvaire's unique contributions to aviation throughout its successful 25-year history. 144 pages, 75 illustrations, 4 full-color pages. Book No. 3618, $12.95 paperback only

PLANES WITHOUT PILOTS: Advances in Unmanned Flight—Bill Siuru

Let 24-year Air Force veteran Siuru show you the state of unmanned flight today. You'll see that future wars could be fought with minimum "face-to-face" contact. Outside the military, unpiloted aircraft can be used for safely testing new aircraft. This insider's look at a technological wavefront will give you new insights to aircraft design. 104 pages, 60 illustrations. Book No. 3432, $12.95 paperback only

BEECHCRAFT STAGGERWING—Peter Berry

Featuring never-before-published facts and photographs from the archives of the Beechcraft Staggerwing Museum Foundation, this book is a complete illustrated record of the colorful history of one of the classic aircraft of the 1930s. Beginning with the early experiences of the Staggerwing's designer, aviation pioneer Walter F. Beech, Berry describes the Staggerwing's development and production, its foreign and domestic uses, and its record-breaking races. This is truly a necessary addition to the library of any classic aircraft buff. 160 pages, 104 illustrations. Book No. 3410, $14.95 paperback only

PAN AMERICAN'S CLIPPERS—Barry Tyler

Relive the historic transoceanic flights of Pan American's Ocean Clippers with this book. You'll find photographs, developmental and service histories, and technical specifications for each model. From the rollout of the first Sikorsky S-40 in 1931 to the retirement of the 40-ton Boeing 314 at the end of World War II, the whole story is here in complete detail. 224 pages, 43 illustrations. Book No. 20002, $15.95 paperback only

THE ERCOUPE—Stanley G. Thomas
Foreword by Fred Weick

This comprehensive, accurate, and personal record chronicles the evolution of the first spin-proof, two-control airplane. You will take a fact-filled trip through the 40-year production history of the Ercoupe, from its conception and development during stall-spin research in 1930, to its groundbreaking jet-assisted takeoff, to the rollout of the final plane in 1970. Photographs, personal recollections, and folklore are included. 144 pages, 61 illustrations. Book No. 20016, $12.95 paperback only

AMERICAN AVIATION: An Illustrated History
—Joe Christy, with contributions by
Alexander T. Wells, Ed.D.

Here, in a comprehensive, well-researched sourcebook, Christy and Wells have taken the history of American aviation and transformed it into a fascinating story of people, machines, and accomplishments that is as entertaining as it is informative. With its hundreds of excellent photographs, this is a book that every aviation enthusiast will want to read and reread! 400 pages, 486 illustrations. Book No. 2497, $25.95 paperback only

HALF A WING, THREE ENGINES AND A PRAYER: B-17s over Germany—Brian D. O'Neill

Experience what it was like to fly a B-17 in combat at the height of World War II! Here, the missions of one bomber crew come alive through excerpts from the wartime diaries of the crew's navigator and copilot. Major portions of the diaries are reproduced and enhanced by photographs and personal recollections from other crew members. 320 pages, illustrated. Book No. 22385, $16.95 paperback only

LEARJETS: The World's Executive Aircraft
—Donald J. Porter

This compelling book traces the history of Learjets, from conception to current production models—with absorbing accounts of how the modern executive jet evolved and the many problems the Lear encountered during production. Plus, you'll find illustrations that depict the aircraft's innovative design, detailed specifications, and performance data for every model built. 128 pages, 32 illustrations. Book No. 2440, $11.95 paperback only

THE MARTIN B-26 MARAUDER—J. K. Havener

Called the "widow maker" because of its alarmingly high accident rate, the B-26 Marauder was almost terminated after a Congressional investigation. This book recounts how this boldly designed aircraft was saved, traces its evolution into a superb bomber, and tells the story of the people who designed, built, and flew it in World War II. 272 pages, 150 illustrations. Book No. 22387, $16.95 paperback only

STEALTH TECHNOLOGY: The Art of Black Magic—J. Jones, edited by Matt Thurber

Revealed in this book are the findings of careful research conducted by an expert in military aviation stealth applications. The profound effect modern stealth technology is having through the United States and world is discussed, along with the facts about the B-2 Stealth Bomber and the F-117A fighter. 160 pages, 87 illustrations, 4-page full-color insert. Book No. 22381, $14.95 paperback only

THE FIRST TO FLY: Aviation's Pioneer Days
—Sherwood Harris

Based on diaries, letters, interviews, and newspaper reports, this book captures the color, adventure, resourcefulness, mechanical ingenuity, and perseverance of the men and women who turned their dreams of flying into reality. Whenever possible, the daring exploits of these fledgling aviators are told in their own words and the words of those who witnessed the great events that marked the early days of aviation, from 1900 to 1915. 240 pages, 72 illustrations. Book No. 3796, $14.95 paperback, $24.95 hardcover

**INSIDE THE STEALTH BOMBER:
The B-2 Story**—Bill Scott, with an introduction by Col. Richard S. Couch, USAF

This book offers an exclusive behind-the-scenes account of the United States' most secret defense project from those who knew it best. Bill Scott brings to light information that until now has been kept from the public eye. He delivers a detailed account of every phase in the bomber's development, as well as a revealing discussion on the technological capabilities and flight characteristics that make the B-2 unique among military aircraft. 240 pages, 42 illustrations. Book No. 3822, $18.95 hardcover only

SWEETWATER GUNSLINGER 201
—Lt. Commander William H. LaBarge, U.S. Navy and Robert Lawrence Holt

Set aboard the U.S. Carrier *Kitty Hawk* during the Iranian crisis of 1980, this fast-paced novel details the lives, loves, dangers, trials, tribulations, and escapades of a group of Tail Hookers (Navy carrier pilots). As much fact as fiction, it's a story that is both powerful and sensitive . . . by authors who do a masterful job of bring the reader aboard a modern aircraft carrier and into the cockpit of an F-14! 192 pages. Book No. 28515, $14.95 hardcover only

Prices Subject to Change Without Notice.

Look for These and Other TAB Books at Your Local Bookstore

To Order Call Toll Free 1-800-822-8158
(in PA, AK, and Canada call 717-794-2191)

or write to TAB Books, Blue Ridge Summit, PA 17294-0840.

Title	Product No.	Quantity	Price

☐ Check or money order made payable to TAB Books

Charge my ☐ VISA ☐ MasterCard ☐ American Express

Acct. No. _____ Exp. _____

Signature: _____

Name: _____

Address: _____

City: _____

State: _____ Zip: _____

Subtotal $ _____

Postage and Handling
($3.00 in U.S., $5.00 outside U.S.) $ _____

Add applicable state and local
sales tax $ _____

TOTAL $ _____

TAB Books catalog free with purchase; otherwise send $1.00 in check or money order and receive $1.00 credit on your next purchase.

Orders outside U.S. must pay with international money order in U.S. dollars.

TAB Guarantee: If for any reason you are not satisfied with the book(s) you order, simply return it (them) within 15 days and receive a full refund.

BC